4th International Conference on Advanced Materials Research and Manufacturing Technologies (AMRMT 2019)

IOP Conference Series: Materials Science and Engineering
Volume 647

Oxford, United Kingdom
8 - 11 August 2019

ISBN: 978-1-7138-0908-1
ISSN: 1757-8981

Printed from e-media with permission by:

Curran Associates, Inc.
57 Morehouse Lane
Red Hook, NY 12571

Some format issues inherent in the e-media version may also appear in this print version.

Institute of Physics
Dirac House, Temple Back
Bristol BS1 6BE UK

Phone: 44 1 17 929 7481
Fax: 44 1 17 920 0979

techtracking@iop.org

Additional copies of this publication are available from:

Curran Associates, Inc.
57 Morehouse Lane
Red Hook, NY 12571 USA
Phone: 845-758-0400
Fax: 845-758-2633
Email: curran@proceedings.com
Web: www.proceedings.com

TABLE OF CONTENTS

MATERIALS SCIENCE

MANUFACTURING TECHNOLOGIES

IOP Conf. Series: Materials Science and Engineering **647** (2019) 011001 doi:10.1088/1757-899X/647/1/011001

PREFACE

2019 4th International Conference on Advanced Materials Research and Manufacturing Technologies (AMRMT 2019) was held in Oxford, UK during Aug. 8-11, 2019. AMRMT 2019 was organized by Hong Kong Society of Mechanical Engineers, supported by South China University of Technology and St Anne's College, University of Oxford. The conference provides a useful and wide platform both for display the latest research and for exchange of research results and thoughts in Materials Research and Manufacturing Technologies and other topics. The participants of the conference were from almost every part of the world, with background of either academia or industry, even well-known enterprise. The success and prosperity of the conference is reflected high level of the papers received.

The proceedings are a compilation of the accepted papers and represent an interesting outcome of the conference. This book covers 2 chapters: Materials Science; Manufacturing Technologies.

We would like to acknowledge all of those who supported AMRMT 2019. Each individual and institutional help were very important for the success of this conference. Especially we would like to thank the organizing committee for their valuable advices in the organization and helpful peer review of the papers.

We sincerely hope that AMRMT 2019 will be a forum for excellent discussions that will put forward new ideas and promote collaborative researches. We are sure that the proceedings will serve as an important research source of references and the knowledge, which will lead to not only scientific and engineering progress but also other new products and processes.

Roderick Smith, Imperial College, UK
Alan Kin-Tak Lau, Swinburne University of Technology, Australia

IOP Conf. Series: Materials Science and Engineering **647** (2019) 011002 doi:10.1088/1757-899X/647/1/011002

COMMITTEES

Conference Co-Chairs

Alan Kin-Tak Lau, Swinburne University of Technology, Australia

Roderick Smith, Imperial College, UK

Conference Program Co-Chairs

Gong Hao, National University of Singapore, Singapore

Ruxu Du, South China University of Tcchnology, China

Advisory Committee

Zhengwei You, Donghua University, China

Yu Dong, Curtin University, Australia

Joong Hee Lee, Chonbuk National university, Korea

Ada Pui Ying Hung, Swinburne University of Technology, Australia

Xiao Feng Li, Shanghai Maritime University, China

Clara Chan, Lee Kee Group, Hong Kong SAR

Sam TS Chin, Hong Kong Aircraft Engineering Company Limited, Hong Kong SAR

Jinsong Leng, Harbin Institute of Technology, China

Thanchanok Ratvijitvech, Mahidol University, Thailand

Nat Vorayos, Chiang Mai University, Thailand

Watchanida Chinpa, Prince of Songkla University, Thailand

Chatchai Putson, Prince of Songkla University, Thailand

Varunee Premanond, King Mongkut's University of Technology Thonburi, Thailand

Hing-Ho Tsang, Swinburne University of Technology, Australia

International Technical Committee

Chamil Abeykoon, University of Manchester, UK

Christopher Cheeseman, Imperial College London, UK

Dan Sun, Queen's University Belfast, UK

Grzegorz Lisak, Nanyang Technological University, Singapore

Maxim Kireitseu, Delft University of Technology, the Netherlands

Mihai Caleap, University of Bristol, UK

N.Ethiraj, Dr. M.G.R Educational and Research Institute, India

Prabhudev M.S, Governement Polytechnic Kalgi, India

Qian Shunzhi, Nanyang Technological University, Singapore

Teik-Thye Lim, Nanyang Technological University, Singapore

Zhongwen Yao, Queen's University, Canada

IOP Conf. Series: Materials Science and Engineering **647** (2019) 011003 doi:10.1088/1757-899X/647/1/011003

Peer review statement

All papers published in this volume of *IOP Conference Series: Materials Science and Engineering* have been peer reviewed through processes administered by the proceedings Editors. Reviews were conducted by expert referees to the professional and scientific standards expected of a proceedings journal published by IOP Publishing.

Thermal Loading Effect During Machining of Borosilicate Glass Using ECDM Process

R K Arya[1,a] **and A Dvivedi**[1]

[1]Mechanical and Industrial Engineering Department, Indian Institute of Technology Roorkee, Roorkee, 247667, India

[a]Email- aryarajendra10iitr@gmail.com

Abstract. Glass has unique properties for micro fluidic, MEMS and optical applications. However, poor machinability of glass materials limit its applications. ECDM process is one of the most utilized micro machining process for fabrication of glass micro products. In ECDM thermal energy of discharges utilizes to provide thermal loading for removal of material. However, controlled thermal loading requires for machining of glass materials owing to its poor thermal conductivity. High thermal stresses may develop at high thermal loading, which leads to uncontrolled facture of glass material. Therefore, a thorough discussion is needed to understand the effect of thermal loading on glass materials. In this study, the experiments are conducted to investigate the thermal loading effect on borosilicate glass during ECDM process. Also, an analytical model is developed to estimate the induced thermal stress during ECDM process. Theses thermal stresses are discussed during drilling of micro hole on work material. After discussion three zones are identified for machining of borosilicate glass using ECDM process, wherein zone-II (i.e. from 44 V to 52 V applied voltage) is identified for safe machining.

1. Introduction

Micro products of glass materials in miniaturization have gained popularity due to its properties like optical transparency, chemical inertness, high hardness and high stability in exposure to sunlight or UV light. MEMS, Lab-on-a-chip, micro fluidics and optical industry are the application area of glass materials [1]. Machining is one of the essential process for fabrication of such products. However, conventional machining of glass material is challenging due to its high hardness, non-conductivity, low thermal conductivity and brittleness. Non-conventional machining processes based on mechanical, chemical, thermal and their hybrid energies have been used for machining of glass materials. Thermal energy based processes are one of the solution to machine these materials. Electrochemical discharge machining (ECDM) process is one of them. The ECDM process can machine nearly all engineering materials irrespective of their properties. The ECDM process includes electrolytic solution, wherein tool electrode (cathode), auxiliary electrode (anode) and work material are immersed. DC power source is used to provide applied voltage to both the electrodes. As the applied voltage increases hydrolysis of electrolytic solution occurs. This results in evolution of H_2 bubbles form tool electrode and O_2 bubbles from auxiliary electrode. The auxiliary electrode has larger surface area, thus the H_2 bubbles easily concentrate around the small surface area of tool electrode and form a gas film. The gas film act as an insulating film, which ceases the current flow in the circuit. This results in generation of high electric field in the gas film. The discharges takes place due to breakdown of gas film under high electric field and produces thermal energy around the tool electrode. This thermal energy utilizes in machining of work material by melting and vaporization. Also, the thermal energy promotes high temperature chemical etching, which assists the removal of work material. Some research work has been done to effective utilization of the produced thermal energy to enhance the performance of ECDM process. Singh and Dvivedi used textured tool electrode, thereby the discharging was uniformly distributed over the machining surface by increasing the nucleation sites on the textured tool electrode [1]. Arya and Dvivedi improved the discharging phenomena during deep micro hole drilling using ECDM process by quantifying and replenishment of vaporised electrolyte inside the machining zone [2]. Saranya and Sankar reduced the loss of thermal energy from sides of tool electrode by using side insulated tool electrode [3]. Pawariya et al. used sonication of tool electrode and calculated the energy channelization index for effective utilization of sonic energy during electrochemical discharge trepanning process [4]. Few others researchers discussed the material removal mechanism of ECDM process by using finite element modelling [5, 6]. Although, the thermal energy utilization has been improved. But, a detailed investigation on thermal loading effect during machining of glass (by considering glass properties) using ECDM process is not reported in existing literatures. The special properties of glass is an important aspect during machining of glass materials using thermal energy. The material removal rate (MRR) is lower at low thermal energy. However, MRR can be improved by increasing thermal energy, but the consequences are high thermal damage and uncontrolled fracture. Hence, a comprehensive investigation is required to understand the effect of thermal loading during machining of glass materials using ECDM process. Therefore, the present article provides a detailed investigation on thermal loading effect during machining of borosilicate glass using ECDM process. An analytical model was developed to calculate induced thermal stress during ECDM process at different applied voltage. A range of applied voltage (in terms of thermal loading) was identified for machining of borosilicate glass using ECDM process.

2. Material and methods

2.1 Experimental setup

The experimental setup is developed in house at a laboratory scale. **Figure 1** is showing the schematic view of fabricated setup of ECDM process. This setup consist of two electrodes (i.e. tool electrode and auxiliary electrode), which are immersed into the electrolyte bath along with the work material.

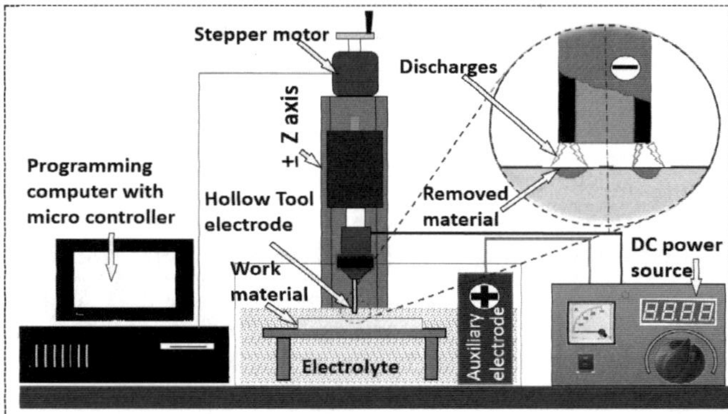

Figure 1. Schematic view ECDM facility with hollow tool electrode

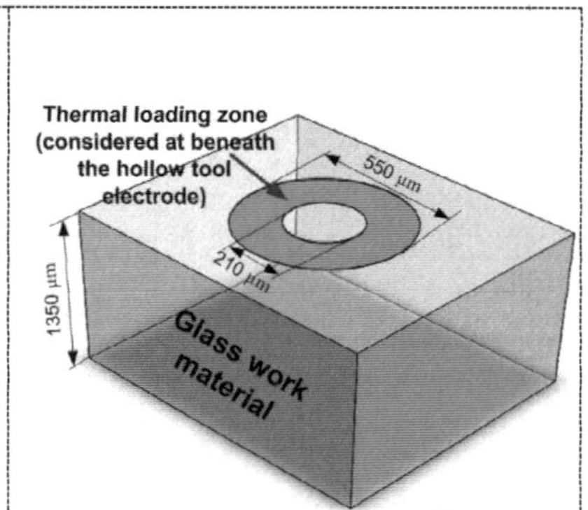

Figure 2. Schematic view of ECDM model assumed for analysis

Borosilicate glass is used as a work material. Both the electrodes are connected to the DC power source, where the tool electrode as a cathode and the auxiliary electrode as an anode. A hollow tool electrode of SS 304 material with 500 μm outer diameter and 250 μm inner diameter is used. The hollow tool electrode provides addition discharging edges and it used in previous ECDM works [2, 7]. The auxiliary electrode of a graphite and the size is 8x5x1 cm^3 is made used. The tool electrode is mounted on a vertical axis (i.e. Z-axis) and the tip of tool electrode is dipped 1-2 mm into the electrolyte bath. The motion of Z-axis is controlled by using a stepper motor with the help of lead screw. The micro controller is employed to control the direction and speed of stepper motor. The micro controller is programmed using a personal computer. The work material is placed under beneath the tool electrode tip. The electrolyte solution used in present investigation is formed by 20 %, wt./ Vol. NaOH aqueous solution.

2.2 Analytical modelling

A model of ECDM process was considered to obtain induced thermal stress in the work material during thermal loading. The model includes a borosilicate glass work material and thermal loading zone (where the thermal energy was supplied from a heat source). The zone was assumed just above the glass surface and beneath the hollow tool electrode. The tool electrode has 500 μm outer diameter and 260 μm inner diameter. The schematic view of assumed model is shown in **Figure 2**.

The assumptions of the model for analysis are as follows:

- It was assumed that the input thermal energy from heat source on interaction plane (i.e. between beneath the tool electrode and above the glass surface) was supplied for small fraction of second (i.e. 6 ms).
- The heat transfer was considered along with the thickness of glass work material.
- The interaction plane was assumed at uniform temperature (i.e. there was no temperature gradient within the plane).
- The model of ECDM process was assumed to be an adiabatic system.
- The model follows Hook's law and produced stresses were considered to be principle stresses along the interaction plane.
- The glass work material was considered at room temperature (i.e. 32 ^{0}C).
- Uniform gas film thickness of 25 μm was assumed around the dipped surface of the hollow tool electrode. Hence, the outer and inner diameter of heat source were 550 μm and 210 μm, respectively.

From the law of conservation of energy;

Net supplied thermal energy to the system = Net thermal energy consumed by the system

= Energy consumed in heating of electrolyte, tool electrode, work material and chemical reactions.

Net supplied thermal energy per unit time 'Q' to the work material during ECDM process is given by,

$$Q = V_a * I_{avg} \tag{1}$$

Net supplied thermal energy 'Q_s' in total duration,

$$Q_s = V_a * I_{avg} * t_{on} \tag{2}$$

where, t_{on} is total heat supplied time (i.e. 6 ms). V_a is Applied voltage (40, 44, 48, 52, 56, 60 volts) and I_{avg} is average current, is given by Jain et al. (1999) [6],

$$I_{avg} = 0.1009 * V_a^{0.4815} * K^{0.3420} * Ø^{0.3420} * d^{0.2881} \tag{3}$$

where, V_a is applied voltage (40-60 V), K is electrolyte conductivity (41.1 mho/m), Ø is diameter of tool electrode (500 μm) and d depth of immersed tool electrode into the electrolyte (2 mm)

As it reported in literature, maximum part (80-90 %) of input energy consumed by electrolyte owing to high thermal conductivity, tool electrode and chemical reactions [4]. Therefore, in this model the heat consumed by work material was assumed to be 14 % of supplied thermal energy.

Percentage of supplied thermal energy to the work material (Q_p) = 0.14* Q_s

$$Q_p = 0.15 * V_a * I_{avg} * t_{on} \tag{4}$$

By using law of conservation, the 'Q_p' is equal to the thermal energy conducted into the work material 'Q_c'. Also, using law of heat conduction, also known as Fourier's law.

Q_c = heat conducted into the work material

$$Q_c = K * A * \left(\frac{\Delta T}{\Delta L}\right) \quad (5)$$

By equating **equation 4** and **5**,

$$0.14 * V_a * I_{avg} * t_{on} = \left[K * A * \left(\frac{\Delta T}{\Delta L}\right)\right]$$

$$\Delta T = \left[(0.14 * V_a * I_{avg} * t_{on}) * \left(\frac{\Delta L}{K * A}\right)\right] \quad (6)$$

where, ΔT is temperature difference in glass work material

Table 1. Process parameters with their nomenclature and their values used in analysis.

S. N.	Parameters name	Nomenclature	Values
1	Applied voltage (V)	V_a	40, 44, 48, 52, 56, 60
2	Total heat supplied time (ms)	t_{on}	6
4	Thermal conductivity of borosilicate glass (W/m-^{0}C)	K	1.2
5	Area of heat supplied on glass surface (m^2)	A	$2.03 * 10^{-07}$
6	Thickness of work material (m)	ΔL	0.0013
7	Poisson ratio of borosilicate glass	μ	.2
8	Young's modulus (GPa)	E	64
9	Linear coefficient of thermal expansion ($^0C^{-1}$)	α	$3.3 * 10^{-06}$

Mathematical expressions for the thermal stress is given by,

$$\sigma_t = \left[E * \alpha * \nabla T * \left(\frac{1}{(1-\mu)}\right)\right] \quad (7)$$

Where, σ_t is induced thermal stress, E is young's modulus of work material, α is linear coefficient of thermal expansion, ΔT is temperature difference in glass work material and μ poisson ratio of work material.

The value of variables used in **Equation 7** are given in **Table 1**.

2.3 Experimental planning

The experiments were performed using ECDM process to investigate the effect of thermal loading on borosilicate glass work material. Applied voltage was varied at (40, 44, 48, 52, 56 and 60) V to control the thermal loading on glass material. Other process parameters were selected from literature and pilot experiments, all are listed in **Table 2**. Further, experiments were conducted to analyse the effect of applied voltage during machining of micro holes on work material. The developed analytical model was used to estimate thermal stresses at same values of applied voltage (those were used in experiments). The experimental and analytical model results were discussed to find out the working range of applied voltage during ECDM process.

Table 2. Process parameters with their values

S. N.	Parameters name	Values
1	Applied voltage (V)	40, 44, 48, 52, 56, 60
2	Electrolyte type / Concentration (%, wt./Vol.)	NaOH / 20
4	Thickness of work material (m)	0.0013

3. Results and discussion

The experimenters were conducted to investigate the effect of thermal loading on work material using ECDM process. The thermal loading was controlled by changing the applied voltage at (40, 44, 48, 52, 56 and 60) V during ECDM process. The other parameters were selected from **Table 2**. The thermal loading was supplied for very short duration 6 ms for all the experiments. Experimental results are plotted in **Figure 3**. As can be seen the thermal damage on work material increased with increase in applied voltage. The reason thereof, as the applied voltage increases the thermal loading increases beneath the tool electrode [2]. This results in high thermal damage on work material. It can be also seen from **Figure 3** the thermal damage was sharply increased after 52 V applied voltage. The microscopic images in **Figure 4** are showing the thermal damage on borosilicate glass with different applied voltages. The thermal damage with micro cracks on work material was almost negligible up to 48 V applied voltage. However, few micro cracks were observed at 52 V applied voltage, but beyond 52 V applied voltage severe thermal damage and micro cracks were observed on glass work material.

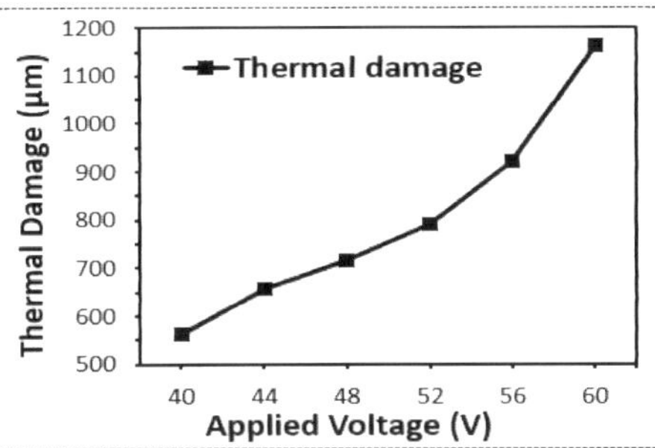

Figure 3. Effect of applied voltage on thermal damage during ECDM process

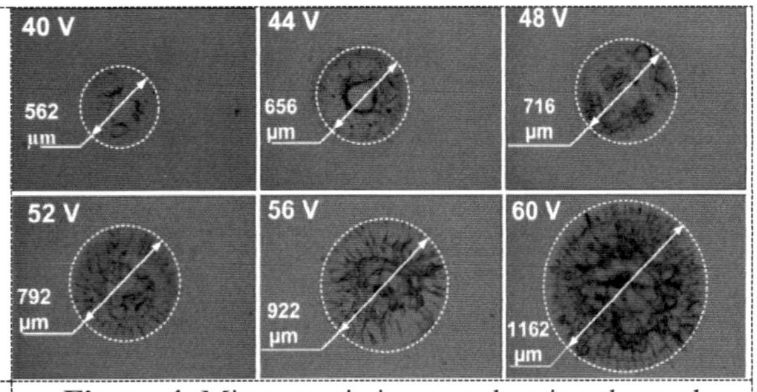

Figure 4. Microscopic images showing thermal damage during ECDM process

Subsequent experiments were conducted to drilling of micro holes on borosilicate glass to observe the machining behaviour of borosilicate glass material using ECDM process at different applied voltages. MRR was calculate for different value of applied voltages (i.e. 40-60 V, with variation of 4V). The MRR was calculated by measuring the weight different of work material in "mg" (i.e. weight of work material before and after machining) divided by machining time in "min". The drilling was performed for 1 min. Experimental results are plotted in secondary axis of **Figure 5**, microscopic images of drilled micro holes are also mentioned**.** It can be seen that the MRR gradually increased till 44 V and similarly after 52 V, but sharp increase in MRR can be observed from 44 V to 52 V applied voltage. This was because, before 44 V applied voltage the thermal loading effect was very low owing to low thermal energy produced at low applied voltage. However, after 52 V applied voltage the MRR slowly increased, because at higher applied voltage the gas film formation process deteriorated during ECDM process [3]. The effective utilization of thermal loading on work material was observed from 44 V to 52 V applied voltage. The effect can also be observed from microscopic images of drilled micro holes. Negligible micro cracks were observed till 52 V applied voltage, after that the high thermal stress resulted in leading of micro cracks to macro cracks or uncontrolled fracture of work material. Hence, thermal stresses were estimated by using **Equation 7** of analytical model and the value of variables are given in the **Table 1**. Same values of applied voltages as in experiments were used to estimate thermal stresses. The results are plotted in primary axis of **Figure 5**. As can be seen, during ECDM process the thermal stresses in the work material increased with increase in applied voltage. The reason thereof is increasing the applied voltage increases thermal loading beneath the tool electrode, which results in increase of sudden change in temperature in the localized region. Also, the glass has poor thermal conductivity, hence the heat generated by thermal loading dissipates very slowly and produces thermal stresses in the localized zone. If, the produced thermal stresses are more than its own tensile strength, micro/macro cracks or uncontrolled fracture initiates on the work material [8].The working range of borosilicate glass were reported 22- 32 MPa [9]. Hence, as the value of induced thermal stresses crosses maximum tensile strength of the work material, it may break or facture at high thermal loading. It can be also seen from **Figure 5,** negligible cracks on periphery of drilled holes were observer from microscopic image even thermal stresses crossed the minimum tensile strength of work material (i.e. till 52 V applied voltage). The reason thereof was the thermal stresses generated only micro cracks and these cracks helped in material removal by thermal spalling [10]. However, after 52 V applied voltage the high thermal stresses resulted in cracks around the periphery of drilled hole and deteriorated the accuracy. Consequently, three zones were identified after the above discussion for machining of borosilicate glass using ECDM process. In zone-I the MRR was very low. Zone-II was identified for high MRR with negligible cracks on periphery of drilled micro hole. Thus, this zone was found best for machining of borosilicate glass using ECDM process. However, in zone-III severe thermal damage or micro/macro cracks were observed, because of high thermal stresses induced at high applied voltage.

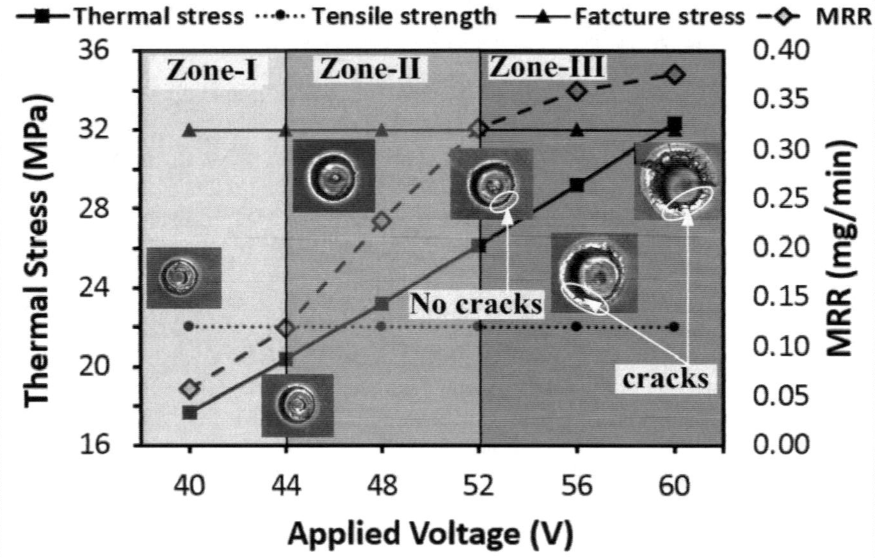

Figure 5. Effect of applied voltage on thermal stress and MRR during machining of borosilicate glass using ECDM process. Also, microscopic images of drilled micro holes at different applied voltage.

4. Conclusion

Thermal loading was controlled by changing the applied voltage. Their effect on borosilicate glass using ECDM process was investigated by performing experiments for very short period of time. An analytical model was developed for estimation of thermal stresses produced during thermal loading from a heat source at different applied voltage. Further, experiments were conducted to investigate the drilling preference of ECDM process at different applied voltage. The drawn conclusions are as;

- Increase in applied voltage increased the MRR, however the thermal damage and thermal stresses also increased.
- During ECDM the induced thermal stresses helped in material removal up to a certain value of applied voltage (i.e. 52 V) without generating cracks on work material.
- Three zones of applied voltage were identified. Zone-II (i.e. from 44- 52 V applied voltage) was found to be the suitable working zone for machining of borosilicate glass using ECDM process.

References

[1] Singh T and Dvivedi A 2018 Journal of Manufacturing Processes **32** 699713
[2] Arya R K and Dvivedi A 2019 Journal of Materials Processing Technology **266** 217229
[3] Saranya S and Sankar AR 2018 Materials and Manufacturing Processes **33** 14221428
[4] Pawariya K, Dvivedi A and Singh T 2019 Precision Engineering **56** 819
[5] Paul L and Korah L.V 2016 Procedia Technology **25** 11751181
[6] Jain V K, Dixit PM and Pandey P M 1999. Int. J. of Machine Tools and Mfg **39(1)** 165-186
[7] Arya R K and Dvivedi A 2018 7th Int. & 28th All India Mfg. Tech. Design and Research (AIMTDR) Conf. (College of Engineering Guindy, Anna University, Chennai, India)
[8] Peitl O and Zanotto E D 1999 Journal of non-crystalline solids **247** 3949
[9] Ashby M F 2012 *Materials and the environment: eco-informed material choice* vol 2, Elsevier
[10] Wuthrich R and Ziki J D A 2014. *Micromachining using electrochemical discharge phenomenon: fundamentals and application of spark assisted chemical engraving,* William Andrew

Improvement in Machined Surface with the use of Powder and Magnetic Field Assisted on Machining Aluminium 6061 Alloy with EDM

Arun Kumar Rouniyar[1] and Pragya Shandilya[1*]

[1]Motilal Nehru National Institute of Technology Allahabad, Pragyagraj-211004, India

*Email: pragya20@mnnit.ac.in

Abstract. Surface characteristics are the major challenges faced by the manufacturing industry when materials are machined with electrical discharge machining. Magnetic Field assisted powder mixed electrical discharge machining (MFAPM-EDM) is hybrid machining process which has shown the good potential of improving the surface characteristics. In this present paper, machining of aluminium 6061 alloy was carried out with aluminium powder mixed in EDM oil and in presence of magnetic field using EDM process Magnetic field, spark on duration, spark off duration, pulse current and powder concentration were selected as variable machining parameters and surface roughness as a response. Box-Behnken design approach based on response surface methodology was used for performing the experiments. In the present analysis, current was observed as the most significant parameter followed by spark on duration, powder concentration and magnetic field on surface roughness. Improvement in surface roughness with lesser voids and craters were observed under the influence of magnetic field and aluminium powder mixed in EDM oil.

1. Introduction

The uses of aluminium 6061 alloy (AA6061) in industrial applications are increasing at present context owing to enticing properties such as light weight, high corrosion resistance, high strength, and better machinability. AA6061 contains magnesium and silicon as principal elements in addition to aluminium. The use of AA6061 was found to be suitable in automobile, marine, aerospace industries and also structural as well as architectural applications [1-2]. Burr formation, adhesion between tool and workpiece, dimensional inaccuracy are some of the problems faced on conventional machining of AA6061 [3, 4]. Owing to certain difficulties faced on conventional machining, researchers have tried to machine the AA6061 by non-conventional machining process. However, electrical discharge machining (EDM) was found suitable which can machine the hard conductive materials irrespective of materials hardness and in the absence of cutting force [5, 6]. In this process, thermal energy is responsible for generating the spark which melts the materials from the surface of workpiece in the form of small debris. At present context, EDM is used in the manufacturing of complex dies, molds and also in aerospace, military, automobile industry for making critical parts. The EDM process, however posses certain limitations such as low machining rate, longer lead time and surface finish limit the use of this process in its applications [7-9]. Thus, researchers have focused their attention on such difficulties by developing new hybrid processes which can improve the process productivity and surface quality. Thus, for effective machining with enhanced machining rate and better surface quality was obtained with the machining of materials with new hybrid machining process known as magnetic field assisted powder mixed electrical discharge machining (MFAPM-EDM). The mechanism of material removal in this process is by melting and evaporation with the assistance of powder in dielectric media and in the presence of magnetic field. The material removal occurs by multiples discharge due to bridging of powder particle under the sparking area instead of single spark combined with Lorentz force developed due to magnetic field coupled electric field which assists in removing material faster to improve the MRR, TWR and better surface integrity [10-12].

Bhatt *et al.* [10] used powder mixed in EDM oil and in presence of magnetic field for machining of AISI D3 and H13 steel. Pulse on time was the most significant parameter for TWR, overcut, microhardness (MH) and surface roughness (SR) followed by powder concentration and peak current. Kant [13] observed higher MRR and lower TWR at 2 g/l of graphite powder mixed in dielectric and copper as tool on machining EN-19 steel using EDM process in presence of fixed magnetic field (0.1 T). Bains *et al.* [14] noticed an improvement in MH, SR and recast layer thickness with the use of SiC powder and magnetic field in EDM on machining Al-Si MMCs. Rouniyar and Shandilya [12] fabricated MFAPM-EDM setup and was successfully used for performing the machining of AA6061. One factor approach was used for experimental design and improvement in TWR and MRR was observed. Hienz *et al.* [15] machined non-magnetic material i.e. Ti-6Al-4V alloy using the EDM process under the influence of external magnetic field. Effect of process parameter on erosion efficiency, debris field and plasma temperature was studied. Rouniyar and Shandilya [16] used Taguchi coupled with GRA to determine the optimum machining parameters on machining Ti-6Al-4V alloy with graphite powder mixed in dielectric for minimum SR and maximum MRR.

Most of the literature on machining of difficult to cut materials were available on machining using either powder mixed in dielectric or magnetic field assisted EDM. Handful works was explored using magnetic field assisted powder mixed EDM on machining of conductive materials. In this paper, the effect of machining parameters on SR, and surface morphology of machined AA6061 with MFAPM-EDM process was studied. Magnetic field (MF), spark on duration (S_{OFF}), spark off duration (S_{ON}), pulse current (I_P) and powder concentration (PC) were selected as variable machining parameters for this study. Second order regression model was developed to predict the SR and surface morphology was analyzed using Scanning Electron microscope (SEM).

2. Experimental Details

Aluminium 6061 alloy (40 x 40 x 10 mm) as workpiece material and electrolytic copper (dia-12 mm) as tool material, EDM oil (grade 5) as dielectric, aluminium (0.074 mm particle size) as powder and NdFeB as ring magnet were selected for this experimental work. Elemental composition (wt %) of Al6061 alloy before machining was measured using EDX test is given as Al (97.87), Mg (0.8), Si (0.4), Ti (0.10), Cr(0.04), Mn (0.12) Fe (0.65), Cu (0.15) and Zn (0.07) [12]. The major properties of Al6061 alloy such as density 2.7 g/cm^3, modulus of elasticity 80 GPa, thermal conductivity 173 W/m K, ultimate tensile strength 270 MPa and hardness 96 BHN respectively. A circular hole of 12 mm diameter was machined on the Al6061 alloy workpiece. Fabricated MFAPM-EDM setup as shown in Figure 1 was used for carrying out experiments. To avoid settlement of powder particles in the tank and for quick removal of debris particle from the spark zone, stirring system and magnetic system were used respectively. Experiments were performed as per Box Behnken design approach of response surface methodology with 5 machining parameters all at three levels. The experimental design consists of total 46 runs (40 factorial points and 6 center point) [17]. The parameters selected for the experimental work were magnetic field, spark on duration, spark off duration, pulse current and powder concentration as illustrated in Table 1. Surface roughness was measured and recorded at three different locations using surface roughness tester with sample length of 4 mm and cutoff length of 0.8 mm and the average of three readings was used for analysis purpose. Surface Morphology at different machining conditions was analyzed using SEM.

IOP Conf. Series: Materials Science and Engineering **647** (2019) 012002 doi:10.1088/1757-899X/647/1/012002

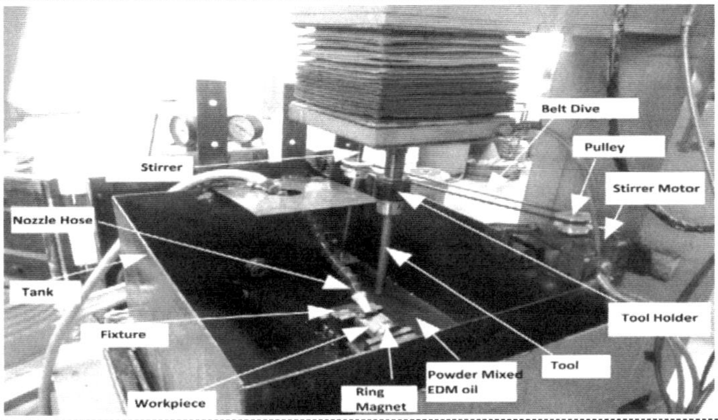

Figure 1. Pictorial View of MFAPM-EDM setup

Table 1. Machining Parameters and Their Levels

Sl. No.	Machining Parameters	Symbol	level 1	2	3	Response	Fixing Criteria
1	Magnetic field (T)	MF	0.15	0.30	0.45	Surface Roughness (SR)	Literature review and Pilot experiments
2	spark on duration (μs)	S_{ON}	30	50	70		
3	spark off duration (μs)	S_{OFF}	35	43	51		
4	Pulse current (A)	I_P	1	4	7		
5	Powder concentration (g/l)	PC	6	8	10		

3. Results and Discussion

Total 46 experiments were performed on machining of AA6061 by MFAPM-EDM process. The average SR observed corresponding to the combinations of experimental runs for machining parameters was used for analysis. Analysis of variance (ANOVA) for SR was carried out at 95 % confidence level using design expert V10.0.1 software as illustrated in Table 2. As observed from Table 2, F-value of 92.45 showing the significance of the developed model. Among linear term I_P, S_{ON}, PC and MF observed were significant as $p<0.05$ was observed. As observed from ANOVA table I_P was most significant parameter which has effect on SR followed by S_{ON}, PC, and MF. Interaction effect was observed for I_P-PC and MF-S_{ON}, while, for quadratic terms significant were observed for I_P, S_{OFF} and PC.

Table 2. ANOVA for SR

Source	SS	dof	MS	F-Value	p-value	% Contribution
Model	40.23	10	4.02	92.45	< 0.0001	
MF	2.53	1	2.53	58.11	< 0.0001	6.52
S_{ON}	6.25	1	6.25	143.52	< 0.0001	16.11
I_P	18.86	1	18.86	433.24	< 0.0001	48.63
PC	3.55	1	3.55	81.54	< 0.0001	9.15
T_{ON}-MF	0.19	1	0.19	4.26	0.0465	0.49
I_P-PC	0.23	1	0.23	5.31	0.0273	0.59
S_{OFF}^2	0.20	1	0.20	4.57	0.0396	0.52
I_P^2	4.98	1	4.98	114.44	< 0.0001	12.84
PC^2	1.91	1	1.91	43.88	< 0.0001	4.92

Residual	1.52	35	0.085			0.22
Lack of Fit	1.32	30	0.044	1.07	0.5258#	
Pure Error	0.20	5	0.041			
Cor Total	41.70	44				
R^2 =0.9635 and Adj. R^2 =0.9531						

SS- Sum of Square, MS- Mean Square, dof-Degree of Freedom, * -Significant, # -Non significant

Using ANOVA, regression coefficient corresponding to linear, quadratic and interaction terms were calculated at 95% confidence level. Thus, second order model to predict the SR was developed in terms of significant machining parameters (actual factors) as illustrated in Equation (1). The pred. R^2 for SR was found to be 0.9222, while Adj. R^2 was observed as 0.9531 and the difference between them was less than 0.2, signifying the relation good agreement of predicted results with the experimental data.

$$SR = +6.47520 + 0.93708 \times MF + 0.052766 \times S_{ON} + 0.67255 \times I_P - 2.15402 \times PC - 0.071750 \times S_{ON} \times MF + 0.040042 \times I_P \times PC - 0.078878 \times I_P^2 + 2.21633 \times 10^{-3} \times S_{OFF}^2 + 0.10990 \times PC^2 \quad (1)$$

Figure 2(a) depicts the surface plot showing the effect of PC and I_P on SR. Increase in SR was observed with increase in I_P and decrease in PC. The increase in SR with increase in I_P was due to increase in energy per pulse in the spark region. A discharge energy associated with high current is capable of removing a chunk of material leading to formation of deep and wide crater with rougher surface. However, decrease in SR with increase in PC was mainly attributed to dispersion of impulsive force among the powder particles. Due to this, spark with less impulsive force strikes the surface with multiple discharges instead of single discharge generating smoother surface. Figure 2(b) illustrates the surface plot showing the effect of MF and S_{ON} on SR. Increase in SR was noticed with increase in S_{ON} and decrease in MF. As S_{ON} increases, higher spark energy is available for machining the workpiece due to longer time duration. Thus, deeper and wide craters were observed leading to poor surface finish. However, decrease in SR with increase in MF was attributed to confinement of plasma causing the increment in depth and reduction in crater size as well as expelling the machining debris from the machining area effectively clearing the gap under sparking area due to Lorentz force developed due to magnetic along with electric field causing proper removal of material leading to lower SR.

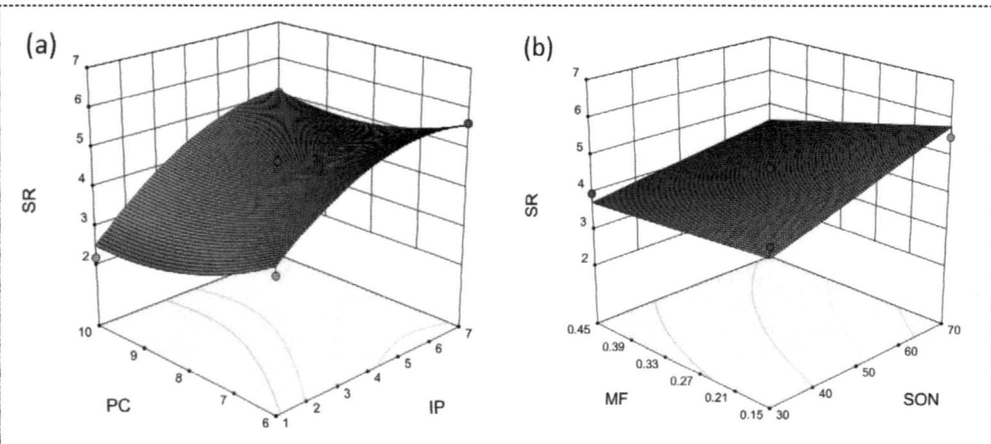

Figure 2. Surface plot of (a) PC and I_P (b) MF and S_{ON} vs. SR

The surface morphology of different machining conditions was analyzed using SEM as shown in Figure 3. Figure 3(a) illustrates the surface morphology of machined sample at higher spark on duration and pulse current (I_P-7A, S_{ON}-70 μs, S_{OFF}-43 μs, PC-8 g/l, MF-0.3 T) where presence of large number of re-solidified debris particles, voids and smaller crater was observed. This was observed due to higher pulse duration available and high pulse current for producing the high discharge energy resulting in higher melting of molten materials from the surface which is unable to flush due to lower pulse off duration resulting in re-solidified layer on the surface. Figure 3(b) depicts the surface morphology at higher magnetic field and powder concentration (I_P-4A, S_{ON}-50 μs, S_{OFF}-43 μs, PC-10 g/l, MF-0.45 T) where the surface appears smoother with presence of lesser voids and re-solidified debris on the machined surface. This was because of effective expulsion of debris at higher magnetic field and due to multiple discharges resulting in shallow crater at higher powder concentration.

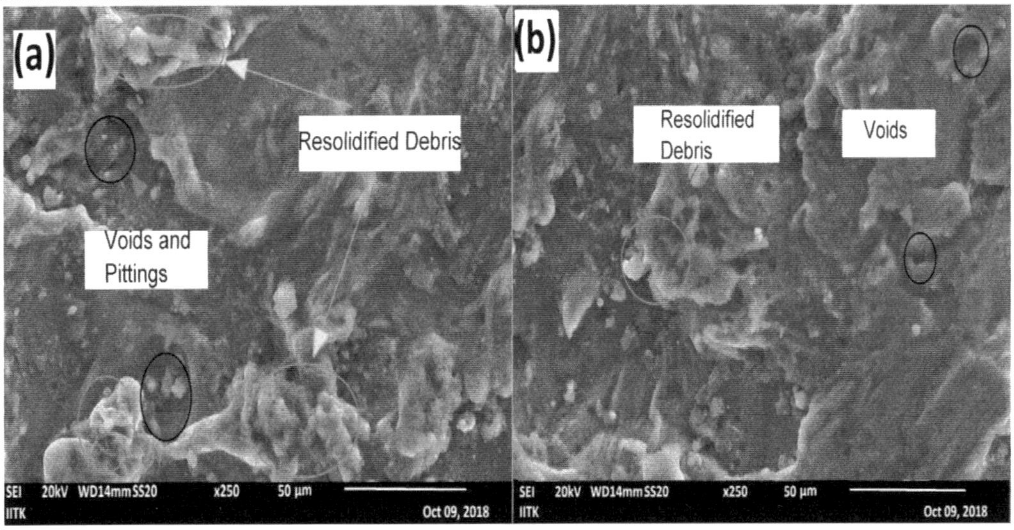

Figure 3. Surface morphology at (a) I_P-7A, S_{ON}-70 μs, S_{OFF}-43 μs, PC-8 g/l, MF-0.3 T (b) I_P-4A, S_{ON}-50 μs, S_{OFF}-43 μs, PC-10 g/l, MF-0.45 T

4. Conclusions

In this study, detailed experimental investigation was carried out to evaluate and identify the effect of machining parameters on surface roughness while machining AA6061 using MFAPM-EDM process with aluminium powder mixed in EDM oil. To check the effectiveness of developed hybrid process scanning electron microscope was used to examine the surface morphology of machined AA6061. Based on experimental observations following findings were derived from the present works.

1. Pulse current was observed as most significant machining on surface roughness at 95% confidence level followed by spark on duration, powder concentration and magnetic field.
2. Second order RSM model has been developed was developed to predict the surface roughness in terms of significant machining parameters.
3. Improvement in surface roughness was observed with the addition of aluminium powder and in presence of magnetic field.
4. SEM image depicts the lesser crack, voids and re-solidified debris on surface when machined at higher magnetic field and powder concentration.

References

[1] Elangovan K, Balasubramanian V, and Valliappan M 2007 Influences of tool pin profile and axial force on the formation of friction stir processing zone in AA6061 aluminium alloy. *Int. J. Adv. Manuf. Technol.* **38** 285–295 Doi:10.1007/s00170-007-1100-2

[2] Rajakumar S, Muralidharan C, and Balasubramanian V 2011 Predicting tensile strength, hardness and corrosion rate of friction stir welded AA6061-T6 aluminium alloy joints. *Mat. Des.*

32(5) 2878-2890.
[3] Sanchez J M, Rubio E, Alvarez M, Sebastian M A and Marcos M 2005 Microstructural characterisation of material adhered over cutting tool in the dry machining of aerospace aluminium alloys. *J. Mat. Process. Technol.* **164–165** 911–918.
[4] Roy P, Sarangi S K, Ghosh A and Chattopadhyay A K 2009 Machinability study of pure aluminium and Al–12% Si alloys against uncoated and coated carbide inserts. *Int. J. Refr. Met. Hard Mater.* **27** 535–544.
[5] Rajesha S, Sharma A K and Kumar P 2012 On electro discharge machining of Inconel 718 with hollow tool. *J. Mater. Eng. Perform.* **21(6)** 882-891
[6] Bisaria H and Shandilya P 2018 The machining characteristics and surface integrity of Ni-rich NiTi shape memory alloy using wire electric discharge machining. *Proc. Inst. Mech. Eng. Part C: J Mech. Eng. Sci.* DOI: 10.1177/0954406218763447.
[7] Pramanik A, Basak A K, Islam M N and Littlefair G 2015 Electrical discharge machining of 6061 aluminium alloy. *Trans. Nonferr. Met. Soc. China* **25(9)** 2866-2874.
[8] Bisaria H and Shandilya P 2017 Experimental studies on electrical discharge wire cutting performance of Ni-rich NiTi shape memory alloy. *Mat. Manuf. Process.* **33(9)** 977-985.
[9] Bisaria H and Shandilya P 2015 Machining of metal matrix composites by EDM and its variants: A Review. *DAAAM International Scientific Book* **33** 267-282.
[10] Bhatt G, Batish A and Bhattacharya A 2015 Experimental investigation of magnetic field assisted powder mixed electric discharge machining. *Part. Sci. Tech.* **33** 1-28.
[11] Kumar H and Davim J P 2011 Role of powder in the machining of Al-10%SiCp metal matrix composites by powder mixed electric discharge machining. *J. Comp. Mat.* **45** 133–151.
[12] Rouniyar A K and Shandilya P 2019 Fabrication and Experimental Investigation of Magnetic Field Assisted Powder Mixed EDM on machining of Aluminum 6061 Alloy. *Proc. IMechE, Part B: J. Engineering Manufacture doi:*
[13] Kant K 2015 Experimental investigation of magnetic field on EDM process by using Taguchi method on EN-19 tool steel. *Glob. J. Res. Eng.: Mech. & Mechan.* **15(3)** 27-35.
[14] Bains P S, Sidhu S S, Payal H S and Kaur S 2018 Magnetic Field Influence on Surface Modifications in Powder Mixed EDM. *Silicon* Doi.org/10.1007/s12633-018-9907-z.
[15] Heinz, K, Kapoor S G, Devor R E and Surla V 2011 An Investigation of magnetic- field-assisted material removal in Micro-EDM for nonmagnetic materials. *J. Manuf. Sci. Eng.* **133** 021002-1-9.
[16] Rouniyar A K and Shandilya P 2018 Multi-Objective Optimization using Taguchi and Grey Relational Analysis on Machining of Ti-6Al-4V Alloy by Powder Mixed EDM Process. *Mat. Today: Proceedings* **5(11)** 23779-23789.
[17] Montgomery D C 1997 *Design and Analysis of Experiments* (New York- Wiley).

IOP Conf. Series: Materials Science and Engineering **647** (2019) 012003 doi:10.1088/1757-899X/647/1/012003

Preparation of Carbon Encapsulated Magnetic FeCo Alloy Nanoparticles Supported on Carbon Nanotubes for Enhanced Microwave Absorption

Guanming Yuan[1,2,a], Yong Su[3], Zhengwei Cui[1,2], Zhijun Dong[1,2], Xuanke Li[1,2,b]

[1] The State Key Laboratory of Refractories and Metallurgy, Wuhan University of Science and Technology, Wuhan 430081, China
[2] Hubei Province Key Laboratory of Coal Conversion & New Carbon Materials, Wuhan University of Science and Technology, Wuhan 430081, China
[3] Shenzhen Tianding New Material Co., Ltd., ShenZhen 518057, China

[a] yuanguanming@wust.edu.cn, [b] xkli@21cn.com

Abstract. A mixture of multi-walled carbon nanotubes in an ethanol solution of $Fe(NO_3)_3.9H_2O$ and $Co(NO_3)_2.6H_2O$, were used to prepare an Fe_2O_3-CoO catalyst supported on carbon nanotubes by a supercritical fluid drying method. The carbon encapsulated FeCo nanoparticles were prepared by the catalytic decomposition of methane over the catalyst at 850°C for 30 mins. The morphology and microstructure of the catalyst and carbon encapsulated nanoparticles (CEP) were characterized by XRD, FESEM, EDS and TEM analyses. The electromagnetic characteristics and microwave absorption properties of the CEP product were also studied. Results show that the CEP product possesses high complex permittivity and good permeability in the microwave frequency range from 2 to 18 GHz. The maximum absorption peak was observed at 9.44 GHz for a large reflection loss (R) of 13.3 dB, with a band width of 7.4 GHz and 2.56 GHz, at R< -5 dB and R< -10 dB, respectively.

1. Introduction

Since S. Gangopadhyay [1] discovered a typical core-shell type of structure with an α-Fe core and an oxide coating, widespread attention has been given to researching metal nanoparticles, which have been encapsulated in protective carbon shells (carbon nanocapsules) [2-8]. These novel structures can induce unique electronic [9,10], optical [11], and magnetic [6,8,12] properties, which are undoubtedly important for the evolution of nanotechnology. Further to this, they can lead to a variety of potential applications, such as magnetic nanoparticles in recording media. This is due to their environmental stability and low magnetic coupling between individual particles, which facilitate the handling of such nanoscale air-sensitive materials [13,14].

With the fast advancement of wireless communication, microwave absorbing materials are becoming increasingly important for specialized outdoor purposes, such as silent rooms, radar systems and military applications. Current electromagnetic wave absorption materials do not display the properties of "strong absorption, low density, wide absorption frequency, thin thickness and compatibility characteristics" [15]. Carbon and carbon nanotubes which have been used to encapsulate magnetic metal particles (i.e., carbon encapsulated nanoparticles, CEP) are one of the new microwave absorbents recently reported [16-20] in an attempt to solve this problem. This is due to their improved electromagnetic characteristics and the protection provided to the encapsulated magnetic nanoparticles against environmental oxidization and chemical erosion.

In this work, the carbon encapsulated iron and cobalt nanoparticles were synthesized by catalytic decomposition of methane over an Fe_2O_3-CoO catalyst supported on carbon nanotubes (Fe_2O_3-CoO/CNTs). The morphologies and microstructure of the catalyst and the CEP product were characterized using XRD, EDS, FESEM and TEM. The dynamic electromagnetic parameters and absorption properties of the CEP product used as microwave absorbing materials were investigated.

2. Experimental

2.1. Preparation of multi-walled carbon nanotubes

The multi-walled carbon nanotubes were synthesized by catalytic decomposition of methane over a NiO/SiO_2 aerogel catalyst [21]. The carbon product was then treated by magnetic stirring in hydrofluoric acid for 48 hours to remove the silica support and the nickel particles. Finally the sample was thoroughly washed in distilled water to remove the hydrofluoric acid.

2.2. Preparation of the catalyst and the CEP

$Fe(NO_3)_3.9H_2O$ and $Co(NO_3)_2.6H_2O$ were mixed at a molar ratio of 66：34 in ethanol using magnetic stirring for 12 hours at room temperature to get an alcosol Then certain quantities of purified carbon nanotubes used as carrier were added into the alcosol to form a homogeneous mixture by using strong magnetic agitation and ultrasonic vibration. The mixture was then placed in an autoclave and dried under the supercritical conditions of 240°C and 7.0 MPa pressure for 15 mins. The alcohol was then slowly discharged and the autoclave was cooled to room temperature with a flowing nitrogen to obtain an Fe_2O_3-CoO/CNTs catalyst.

About 1 g of the catalyst was placed into a quartz boat and then inserted into a tubular furnace, then heated at a rate of 5 °C/min under nitrogen to 850°C. The nitrogen gas was replaced by hydrogen gas and then held for 30 mins in order to reduce the catalyst. Thereafter a mixture of methane and hydrogen with a flowing ratio 200/100 in mL/min was introduced to promote the decomposition reaction for 30 mins. After the reaction, the product was cooled to room temperature under a nitrogen atmosphere to obtain the CEP product.

2.3. Characterization

X-ray diffraction (XRD) analysis of the catalyst and the carbon product were performed by a D\MAXIIIc X-ray diffractometer using filtered Cu radiation. The morphologies and microstructure of the carbon nanotubes, the catalyst and the CEP product were characterized by a Carl Zeiss LEO 60 mm 1530 field emission gun scanning electron microscope (FESEM) equipped with an energy dispersive spectroscopy (EDS) and transmission electron microscopy (TEM) was conducted on a JEOL JEM-2000FX II.

3. Results and discussion

3.1. Morphology and structure of the carbon nanotubes, the catalyst and the CEP

FESEM and TEM images of purified multi-walled carbon nanotubes used as metal catalyst carrier are shown in Figure 1. It shows that the carbon nanotubes are curved and possess a diameter of 20~40 nm and a length typically 3~8 μm. Figure 2(a-b) shows the FESEM and TEM images of the typical catalyst, which mainly consists of nanosized particles supported on carbon nanotubes. The interfacial adhesion strength between the nanoparticles and the carbon nanotubes is strong, despite a few particles peeling off from the walls of carbon nanotubes. It can be seen from Figure 2(c-d) that the CEP product mainly consists of particles with a diameter in the range 50~100 nm and the carbon nanotubes are covered up by the particles.

IOP Conf. Series: Materials Science and Engineering **647** (2019) 012003 doi:10.1088/1757-899X/647/1/012003

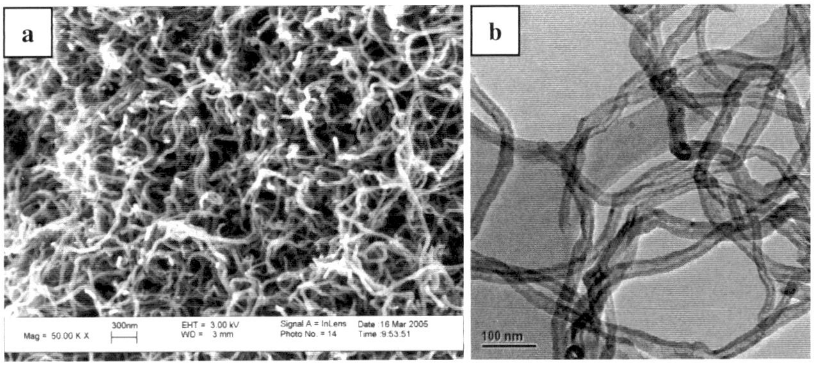

Figure 1. Typical (a) FESEM and (b) TEM images of purified multi-walled carbon nanotubes.

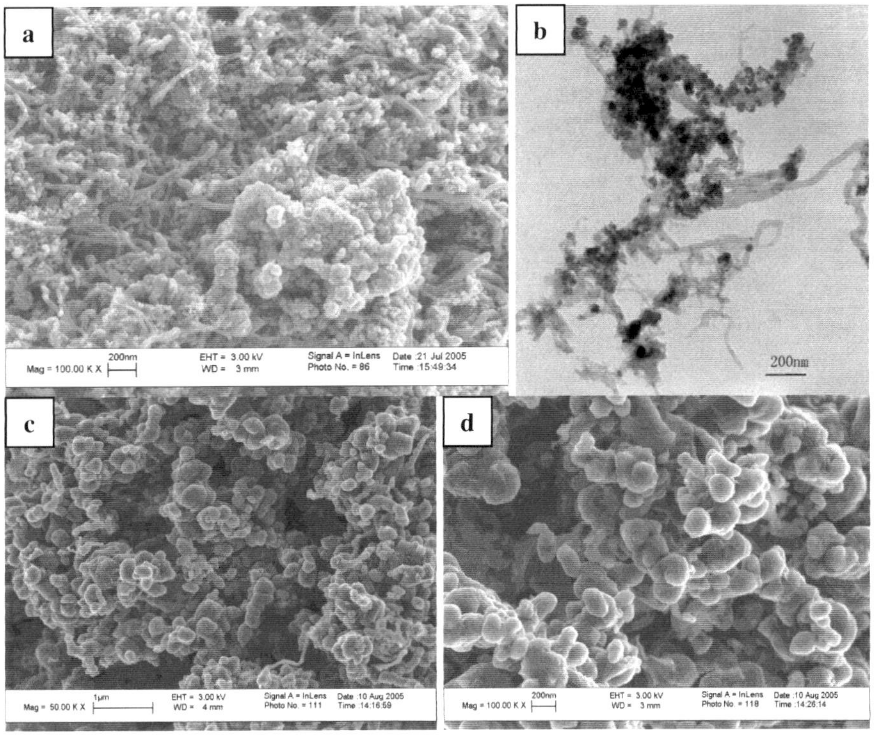

Figure 2. Typical (a) FESEM, (b) TEM images of the catalyst and (c-d) FESEM of the CEP product.

Figure 3 shows the XRD patterns of the catalyst and the CEP product obtained at 850°C for 30 mins, which illustrates that the catalyst mainly consists of carbon, $CoFe_2O_4$, and Fe_2O_3. The CEP product mainly consists of carbon and a Co_3Fe_7 alloy, with a small presence of an $Fe_{64}Ni_{36}$ alloy as a by-product. This by-product can be attributed to the carbon nanotubes that have encapsulated a few Ni particles [21] and are not completely eliminated in this purification process.

IOP Conf. Series: Materials Science and Engineering **647** (2019) 012003 doi:10.1088/1757-899X/647/1/012003

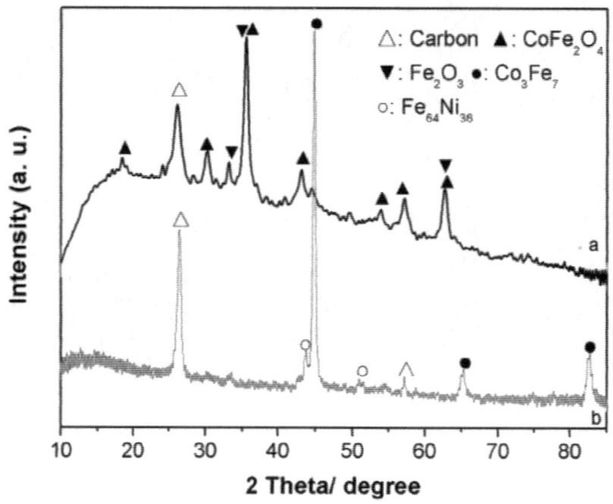

Figure 3. XRD patterns of the catalyst (a) and the CEP product (b).

EDS spectrum of the catalyst and the CEP product are shown in Figure 4, which indicates that the catalyst and the CEP product are both mainly composed of three elements, i.e., carbon, iron and cobalt. The elemental content of iron in the catalyst was higher than that found in the CEP product, in contrast, the carbon content for the catalyst was lower than that of the CEP product. The elemental weight ratio of the CEP product listed in Table 1 further confirms the above result. The EDS analysis also discovered a small amount of nickel, which is in agreement with the XRD pattern analysis.

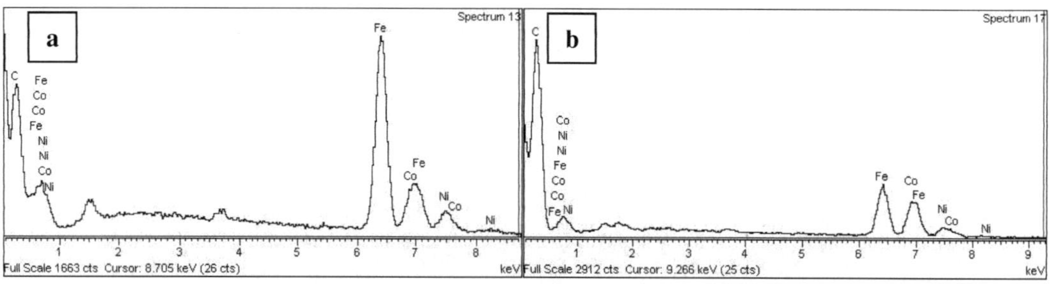

Figure 4. EDS spectrum of the catalyst (a) and the CEP product (b).

Table 1 The elemental weight ratio of the catalyst and CEP product.

Sample	Elemental weight ratio/ %				
	C	Fe	Co	Ni	X
catalyst	50.18	37.23	7.12	5.00	0.47
carbon product	69.88	16.21	12.17	1.30	0.44

Note: "X" indicates a small mass impurity in the sample.

In order to clearly observe the morphology and microstructure of the CEP product, TEM measurements of these samples are conducted. It can be seen from Figure 5 that the CEP product is linked together and mainly consists of particles with a diameter of 50~100 nm. Figure 5(b) looks like a pearl necklace and the ellipsoidal shaped metal nanoparticles look like the necklace's beads which are completely encapsulated by carbon layers. The thickness of the coating layers observed is about 15~30 nm as shown in Figure 5(c).

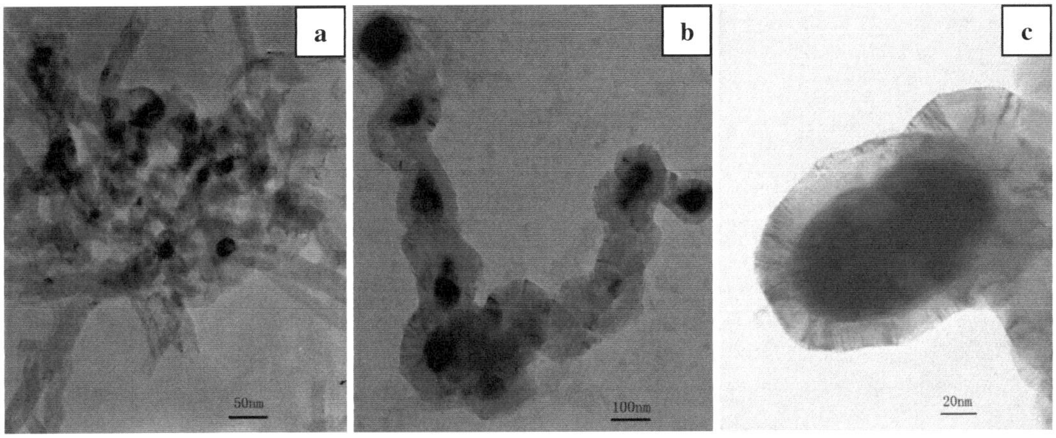

Figure 5. (a-c) Typical TEM images of the CEP product.

The secondary electron image and EDS spectrum distribution of the CEP's are shown in Figure 6. When the electron beam was scanned across two carbon encapsulated particles, it can be observed that the nanoparticles consist of carbon, cobalt and iron. The carbon element was distributed outside the nanoparticles and the metals appear inside, which indicates that the FeCo alloy was completely encapsulated by the carbon, despite potential influence of the thickness of encapsulating layers. The element weight content of nanoparticles by EDS analysis shown in Figure 6(c) is in good agreement with that of CEP product shown in Figure 4(b).

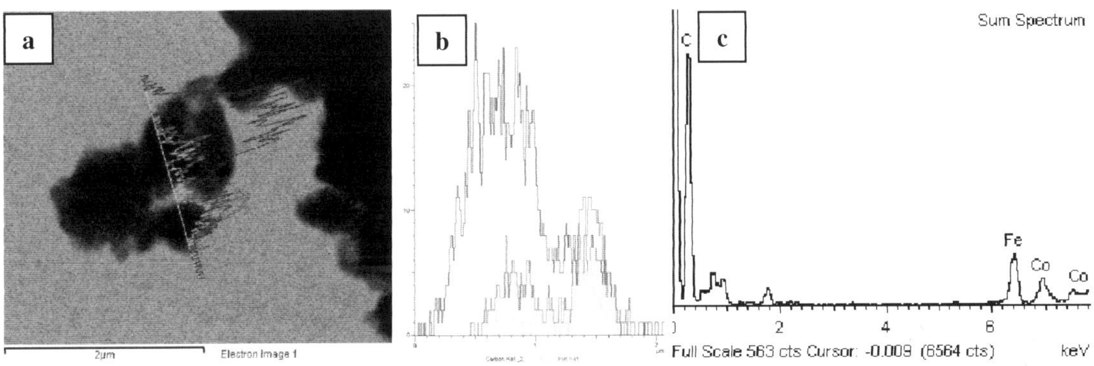

Figure 6. (a) Secondary electron image of the CEP, (b) the normalized intensity variations of the carbon (blue lines), cobalt (green lines) and iron (read lines) edges as the electron beam is scanned across the CEP and (c) EDS spectrum of the CEP.

3.2. Dynamic electromagnetic parameters and absorption property of the CEP

Figure 7 shows the dynamic electromagnetic parameters (complex permittivity $\varepsilon_r'' = \varepsilon' - j\varepsilon''$ and complex permeability $\mu_r'' = \mu' - j\mu''$) of the CEP product's obtained by coaxial measurement, from 2 to 18 GHz, using an S parameter vector network analyzing instrument. Over the range of microwave frequencies 2 to 18 GHz, the change of the real part (ε_r', μ_r') and imaginary part (ε_r'', μ_r'') of complex permittivity are obvious. By increasing the microwave frequency from 2 to 18 GHz, ε_r' decreases from 106.4 to 2.7, and ε_r'' decreases from 240.9 to 16.2. In addition, μ_r' decreases from 1.2 to 0.5 and μ_r'' increases from 0.34 to 1.15, then decreases to 0.2, which indicates that the CEP product possesses good frequency-responding characteristics [22,23].

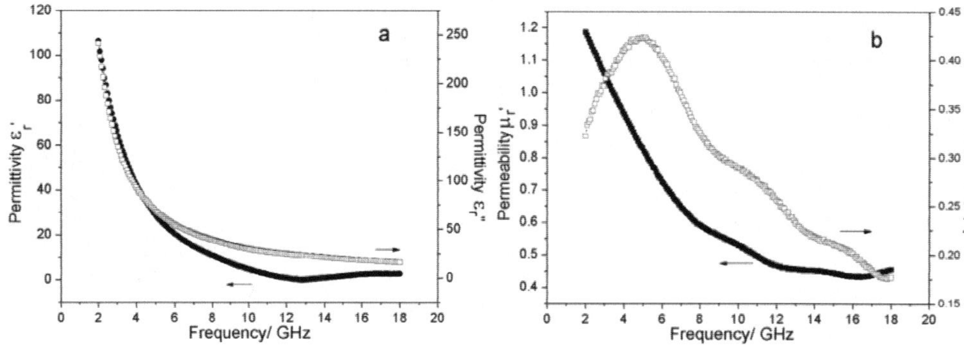

Figure 7. Complex permittivity (a) and complex permeability (b) spectra of CEP product at the microwave frequency range from 2 to 18 GHz.

Figure 8 displays the reflection loss (R) and the frequency curves of olefin composites incorporated with the CEP product at a weight ratio of 1:2, the thickness of the composites is 2 mm. In the microwave frequency range from 2 to 18 GHz, the maximum absorption peak of the composites was 13.3 dB at 9.44 GHz. The composites showed band widths of 7.4 GHz (R< -5 dB) and 2.56 GHz (R< -10 dB) which indicates that the CEP product possess good microwave absorption properties [20,23,24].

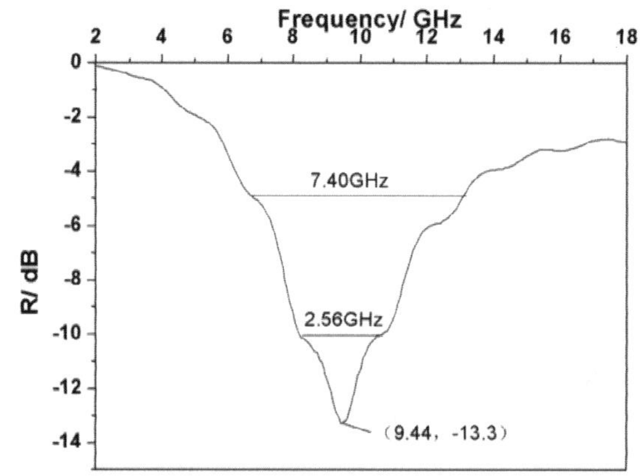

Figure 8. Reflection loss and frequency curves of olefin composites incorporated with (FeCo)@C/CNTs magnetic nanoparticles.

4. Conclusions

Synthesis of carbon encapsulated FeCo magnetic nanoparticles can be controlled by a catalytic chemical vapor deposition method, using an Fe_2O_3-CoO/CNTs catalyst. The dynamic electromagnetic parameters of the prepared encapsulated product show a high complex permittivity and good permeability in the microwave frequency range from 2 to 18 GHz. The prepared carbon encapsulated nanoparticles are good candidates for microwave absorption materials with widely applications as they possess high dielectric loss and ideal microwave absorption properties.

Acknowledgments

This work was sponsored by the Key Laboratory of Hubei Province for Coal Conversion and New Carbon Materials (grant No. WKDM201701) and the China Scholarship Council Fund (201808420114).

References

[1] Gangopadhyay S and Hadjipanayis G C 1992 *Phys. Rev. B* **45** 9778.

[2] Saito Y, Okuda M and Yoshikawa T 1994 *J. Phys. Chem.* **98** 6696.
[3] Saito Y 1995 *Carbon* **33** 979.
[4] Liu B H, Ding J, Zhong Z Y, Dong Z L, White T and Lin J Y 2002 *Chem. Phys. Lett.* **358** 96.
[5] Huo J P, Song H H and Chen X H, 2004 *Carbon* **42** 3177.
[6] Koltypin Y and Palacio F 2004 *Acta Mater.* **52** 2165.
[7] Lu Y, Zhu Z P and Li Z Y 2005 *Carbon* **43** 369.
[8] Paraskevas I, Caps V and Tsang S C 2006 *Carbon* **44** 820.
[9] Pascual J I, Méndez J, Gómez-Herrero J, Baró A M, García N, Landman U, Luedtke W D. Bogachek E N and Cheng H P 1995 *Science* **267** 1793.
[10] Collins P G, Zettl A, Bando H, Thess A and Smalley R E 1997 *Science* **278** 100.
[11] Bubke K, Gnewuch H, Hempstead M, Hammer J and Green M L H 1997 *Appl. Phys. Lett.* **71** 1906.
[12] Dong X L, Zhang Z D, Jin S R and Kim B K 2000 *J. Magn. Magn. Mater.* **210** 143.
[13] McHenry M E, Majetich S A and Artman J O 1994 *Phys. Rev. B* **49** 11358.
[14] Hayashi T, Hirono S, Tomita T and Umemura S 1996 *Nature* **381** 772.
[15] Mu G H, Shen H G, Qiu J X and Gu M Y 2006 *Appl. Surf. Sci.* **253** 2278.
[16] Zou Y H, Liu H B, Yang L and Chen Z Z 2006 *J. Magn. Magn. Mater.* **302** 343.
[17] Fan Z J, Luo G H, Zhang Z F, Zhou L and Wei F 2006 *Mater. Sci. Eng. B* **132** 85.
[18] Zhang J, Müller J O, Zheng W Q, Wang D, Su D S and Schlögl R 2008 *Nano Lett.* **8** 2738
[19] Hisada D, Fujiwar Y, Sato H, Jimbo M, Kobayashi T and Hata K 2011 *J. Magn. Magn. Mater.* **323** 3184.
[20] Afghahia S S S and Shokuhfar A 2014 *J. Magn. Magn. Mater.* **370** 37.
[21] Li X K, Yuan G M, Brown A, Westwood A, Brydson R and Rand B 2006 *Carbon* **44** 1699.
[22] Wu A B, Yang X W, and Yang H 2013 *Dalton Trans.* **42** 4978.
[23] Xu M H, Zhong W, Wang Z H, Au C T and Du Y W 2013 *Physica E* **52** 14.
[24] Zhang D F, Xu F X, Lin J, Yang Z D and Zhang M 2014 *Carbon* **80** 103.

IOP Conf. Series: Materials Science and Engineering **647** (2019) 012004 doi:10.1088/1757-899X/647/1/012004

Using Deep Neural Networks to Predict the Tensile Property of Ceramic Matrix Composites Based on Incomplete Small Dataset

GAO Xiang[1], LI Guanghui[1], TAN Rong[2]and YAO Leijiang[3]

[1] School of Computer Science, Northwestern Polytechnical University, Xi'an 710129, China
[2] School of Software, Northwestern Polytechnical University, Xi'an 710072, China
[3] School of Laboratory of Science and Technology onUAV, Northwestern Polytechnical University, Xi'an 710072, China

sxlllslgh@mail.nwpu.edu.cn

Abstract. It has been a hot spot in the area of material science that how to design an experiment to produce a kind of ceramic matrix composites (CMCs) which possesses ideal performances. An approach is to build models with data from previous experiments recorded in published papers and predict the experiment parameters needed by the ideal CMCs. According to the database of CMCs funded by the National Material Genome Engineering, 8 factors were considered to affect the tensile property of CMCs, which were the basis, the reinforcement fiber type, the reinforcement fiber volume content, the perform type, the porosity, the interface type, the interface thickness and the density. Among the data we collected from papers, however, only few of pieces contained all the 8 factors, most were incomplete, some of them even lacked multiple factors. This paper's work mainly researched how to take advantage of the incomplete data to build an effective model to predict the tensile property of CMCs. We proposed a model to predict the tensile property of CMCs based on a 1-D convolution neural network (CNN), the training data of which were all from papers. To decrease the influences of the incompleteness of data, we tried several methods to process the missing data, such as the mean imputation, the K-Means clustering imputation, the Hot-Deck Imputation and the regression imputation. The results showed that the regression imputation with a dual-hidden-layer feedforward network performed better and improved the performance of the CNN tensile property prediction model.

1. Introduction

In the traditional material science engineering, it used to take decades for experimenting to develop a new material. Further, the conditions like raw materials and environment of experiments depend on the experience and intuition of scientists deeply. As a consequence, the development of new materials progress as a slow pace. When it comes to the 21st century, all walks of life have been developing more and more rapidly, and the demand of various materials which could adapt to different special environments is becoming more and more urgent, so the slow traditional development progress of materials becomes a weak link of the upgrading of industry. Therefore, it becomes a problem to be solved urgently that finding an approach to predict the manufacturing process of materials of which performance meets requirements according to existing ones.

Ceramic matrix composites (CMCs) are a subgroup of composite materials, which consist of various fibres embedded in a ceramic matrix. They not only possess the exceptional properties of ceramic such as the resistance to high temperature, the high strength and rigidity, the low density and the resistance of corrosion, but also improve the disadvantages like poor toughness and reliability. CMCs have great potential and have been applied in the areas of aeronautics, astronautics, automotive industry, et al.[1]

In June, 2011, the United States government published Advanced Manufacturing Partnership (AMP), a crucial part of which is Materials Genome Initiative (MGI)[2]. MGI is composed of computational tools, experimental tools and the digital data. The digital data contains all kinds of data and information from posted experiments, which is used to provide the basis of design for high throughput experiments.

Due to the development process of materials is extremely sophisticated, time-consuming and costly, it is an effective way to extract the information from published documents for collecting research data as much as possible, which could be used to model to provide clues for designing experiments. Owing to the variety of the quality of documents, the lack of experiment factors exists in a certain number of documents, which results in the impossibility to train and predict using existing common models. Hence it is requisite to process the missing data.

In the traditional data process, dropping missing values, zero imputation, mean imputation and Hot Deck Imputation are usually used to process missing data[3]. Furthermore, K-Means clustering imputation and regression imputation based on the machine learning are often used[4-6]. This work compared and tested incomplete data processing methods above, researches the factors influencing the tensile property of CMCs, and builds a model to predict the tensile property of CMCs.

2. The tensile property prediction model

2.1. Data collection and preprocessing

The factors influencing the tensile property of CMCs mainly consist of 8 parts, which are the basis, the reinforcement fibre type, the reinforcement fibre volume content, the perform type, the porosity, the interface type, the interface thickness and the density[7], and we used the tensile strength to measure the tensile property of a material.

We collected 342 pieces of data totally, 67 pieces of which are complete, while there are different degrees of deficiencies in the other 275 ones, the statistics of which is as table 1.

Table 1. The statistic of missing data.

Missing Item	Reinforcement Volume Content	Interface Type	Interface Thickness	Porosity	Density
Count	76	48	106	250	227

As table 1 shows, there is a serious lack of data. In the 275 pieces of data, there are 21 pieces lacking of all the 5 factors mentioned above at the same time.

As the basis, the reinforcement fibre type, the perform type and the interface type are all text data, they could not be processed with mathematic models. So, they were encoded by One-Hot Encoding[8].

As for the other factors, which are the reinforcement fibre type, the porosity, the interface thickness and the density, and the tensile property that is to predict are all digital data. They were scaled between 0 and 1 aiming to improve the performance using neural networks[9].

2.2. Data mining model

This work tried to build the prediction model with a support vector machine (SVM)[10], a deep feedforward network and a 1-D convolution neural network and compared them. The deep feedforward network (DNN), which is also as known as the multilayer perceptron (MLP), is a kind of function approximation model inspired by the neuroscience[11]. and is the most classical neural network model[12]. The SVM model usually performs well on the small dataset, so we used its result to asses the performance of neural networks.

A deep feedforward network consists of several layers, and each layer is composed of several units operating in parallel. Every unit represents a function from a vector to a scalar, which is called a neuron as figure 1 shows.

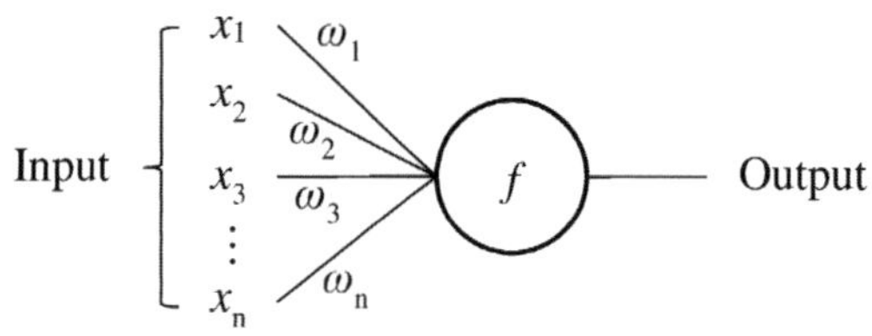

Figure 1. A neuron.

The expression of the function is

$$f(\boldsymbol{x}; \boldsymbol{W}, \boldsymbol{b}) = g\left(\boldsymbol{W}^{\mathrm{T}}\boldsymbol{x} + \boldsymbol{b}\right) \tag{1}$$

$\boldsymbol{x}$ means the vector inputted into the layer, which is

$$\boldsymbol{x} = \begin{pmatrix} x_1 \\ x_2 \\ x_3 \\ \vdots \\ x_n \end{pmatrix} \tag{2}$$

$\boldsymbol{W}$ means the weight matrix, which is

$$\boldsymbol{W} = \begin{pmatrix} \omega_1 \\ \omega_2 \\ \omega_3 \\ \vdots \\ \omega_n \end{pmatrix} \tag{3}$$

$\boldsymbol{b}$ means the bias matrix. In another word, each input $\omega_i x_i$, will be added with a bias b_i, and the output will be calculated with the activation function $g(z)$.

The neural network composed of such neurons is called forward, for there is no feedback connection between the output of the model and the model itself, which is called the feedforward neural network. The vector composed of input values in Figure 1 is generally called the input layer, and the vector composed of neurons in the last layer is usually called the output layer. Other layers between them are often habitually called the hidden layers. If the count of hidden layers is greater than or equal to two, the neural network is often called the deep feedforward network.

A convolution neural network (CNN) is to replace the matrix multiplication of at least one layer in a DNN into the convolution operation. CNN is of sparse interaction, which reduces computation and improves the computation efficiency that could describe the complex interaction among multiple variables[13].

We inputted 8 factors above into the neural networks and let the tensile property be the output. The adaptive moment estimation (Adam)[14] is used as the optimization function, and the rectified linear unit (ReLU)[15] is used to be the activation function. While comparing the performance of DNN and CNN, we filled the missing values with zero, for the neural networks is not sensitive with zero inputs, which means the missing data was not processed[16]. The 90 percent of the dataset was set to the training set, the other 10 percent was as the test set. We used the two neural networks respectively to train models and compared the performance of them on the test set with the root-mean-square error (RMSE) and the coefficient of determination as the evaluation indexes. RMSE is also the loss function in the training of the neural networks. The coefficient of determination, denoted R^2 and pronounced "R squared", is a scalar ranging from 0 to 1. The more R^2 is closed to 1, the better the model fits the data. It is one of the most significant indexes in the regression fitting[17, 18].

2.2.1. Deep feedforward network. We used a dual-hidden-layer feedforward network to build the prediction model, the structure of which is as figure 2 shows.

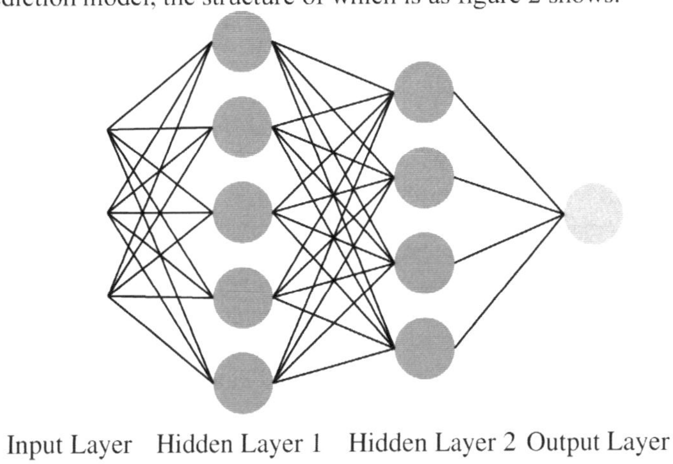

Figure 2. A deep feedforward network with two hidden layers. The number of the neurons in the input layer is 8 that is equal to the dimensions of the input data. The number of the neurons in the output layer is 1, which corresponds the tensile property to predict. The numbers of the neurons in the hidden layers could not be determined by a reliable theory at present, so they should be tested in the experiments below.

2.2.2. 1-D convolution neural network. Given the excellent performance of CNNs while extracting local features, this work also implemented a 1-D convolution neural network, the structure of which is as figure 3 shows.

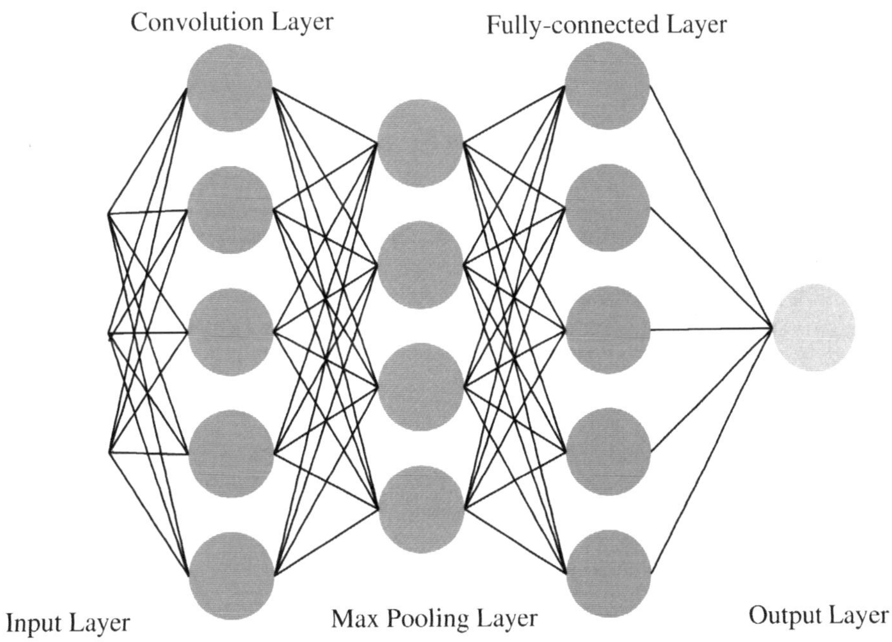

Figure 3. A 1-D convolution neural network.

2.3. Process of missing data

Both neural networks above ignored missing data. For better prediction performance, we tried different methods to process missing data.

In the traditional methods processing missing data, there are roughly two types, which are dropping and filling. In filling methods, there are mean imputation, forward imputation, back imputation, Hot-Deck Imputation and more. With the machine learning developing, there are imputation methods based on K-Means clustering and neural networks appeared. In this work, we tried dropping rows with missing values, mean imputation, Hot-Deck Imputation, K-Means clustering imputation and regression imputation based on the neural network to process missing data, and compared the performance of them.

It is the easiest operation to drop rows with missing values, which could be free from dealing with missing data, but lose information in the part of rows not lost. This method is often used in the situation that there are quite limited influences after parts of data dropped. The mean imputation is to replace any missing value with the mean of that variable for all other cases, which performs well when the values distribute around their mean values. Hot-Deck Imputation is to fill the missing dimensions of a piece of data with values from the most similar but complete piece with it.

K-Means clustering is a kind of unsupervised clustering algorithms. It needs to be assigned the number of clusters, denoted K. It is to divide all objects into K clusters, making every object be nearest to the centre of the cluster it belongs to. K-Means imputation clusters all complete pieces of data firstly, then contrasts every incomplete piece of data with the centres of clusters by their distances. A piece of incomplete data would be filling with the centre, which is generally the mean, of the cluster nearest to it.

As for the regression imputation, is a common imputation method in the traditional statistics that fits known data with a specified function to predict missing data. The fitting function is normally a linear function of a polynomial function, which performs not well when the data is extremely complex. Therefore, we used the regression imputation with neural networks as fitting functions to fill the incomplete data, and then use the neural networks above to predict the tensile property.

3. Result and discussion

3.1. Comparison of the models

The SVM model, usually called the support vector regression (SVR)[19] model, possesses a large number of hyperparameters, two of which often need to be tested by the grid search, the gamma and the penalty parameter C. We also used 10-fold cross validation to prevent overfitting. When the C is 5.0 and the gamma is 0.1, we got the best R^2 score on the test dataset, which is 0.8935.

Aiming to find the best numbers of neurons of hidden layers in the feedforward neural network, we tested different numbers of neurons by the grid search method. The R^2 scores of the models on the test set after 500 epochs are as table 2 shows.

Table 2. The R^2 scores on the test dataset with different numbers of neurons of hidden layer after 500 epochs.

2nd Layer \ 1st Layer	20	40	60	80	100	120	140	160
20	0.9532	0.9764	0.9665	Nan	0.9906	Nan	0.9715	0.9892
40	0.9876	0.9781	Nan	0.9925	Nan	0.9682	0.9942	Nan
60	0.9914	0.9821	Nan	0.9819	0.9517	0.9831	0.9951	0.9779
80	0.9569	Nan	Nan	0.9914	0.9950	0.9810	0.9920	Nan
100	0.8950	0.9613	0.9915	0.9061	0.9941	0.9636	0.9764	Nan
120	0.8962	0.9832	0.9363	0.9720	0.9939	0.9870	0.9804	0.9809
140	0.9650	0.9711	0.9602	0.9738	0.9784	Nan	0.9918	0.9626
160	0.9840	Nan	0.9542	0.9719	0.9898	0.9867	0.9861	0.9870

According to table 2, the R^2 scores after 500 epochs could reach a range from 0.8950 to 0.9951. Some R^2 scores are Nan, for their models does not converge. The variance of R^2 scores in table 2 is 0.000527.

For finding the best numbers of the neurons of the convolution layer (also as known as the convolution kernels or the filters) and the fully-connected layer, we set the size of the convolution kernels to 5, and tested different numbers of the filters and the neurons of the fully-connected layer by the grid method. The R^2 scores of the models on the test dataset after 500 epochs are as table 3 shows.

Table 3. The R^2 scores with different numbers of filters and neurons of the fully-connected layer after 500 epochs.

Fully Connected Layer \ Filters	20	40	60	80	100	120	140	160	180
20	0.9975	Nan	0.9915	0.9966	Nan	0.9950	0.9971	Nan	0.9900
40	0.9961	0.9800	0.9954	Nan	0.9900	Nan	0.9961	0.9783	Nan
60	0.9914	0.9897	0.9812	0.9898	0.9454	0.9952	0.9954	0.9837	0.9534
80	Nan	0.9764	0.9636	0.9787	0.9961	0.9953	0.9941	0.9989	0.9884
100	0.9907	0.9871	0.9920	0.9877	Nan	Nan	Nan	0.9970	0.9934
120	Nan	0.9952	0.9707	0.9950	Nan	Nan	0.9934	0.9883	0.9639
140	0.9931	0.9952	0.9948	0.9930	0.9969	0.9956	0.9912	0.9662	0.9810

As is shown in table 3, the R^2 scores could reach a range from 0.9454 to 0.9989 after 500 epochs, which means the CNN performs better than the tow-layer feedforward neural network. Furthermore, the variance of R^2 scores of CNN is 0.000141, which is nearly one fourth of that of DNN.

Therefore, we used the CNN as the model to train and predict the tensile property of CMCs. The size of the filters was set to 5, and the number of which was set to 160. The number of the neurons of the fully-connected layer was set to 80. Eventually, the R^2 score could reach 0.9989.

3.2. Comparison of different imputation methods

In section 3.1, we set all missing values to zero, which means we let the neural networks ignore the missing data. In order to achieve a better performance to predict the tensile property, we tried several different methods to process the missing data. The model to predict the tensile property is the CNN model above.

3.2.1. Dropping the rows with missing values. All the 275 pieces of data with missing values was dropped. In the 67 pieces of data left, 90% of them was as the training set, and 10% of them was as the test set. After 500 epochs, the R^2 score was only 0.9880.

3.2.2. Mean imputation. The R^2 score was 0.9971.

3.2.3. Hot-Deck Imputation. There is no strict definition for the similarity among the pieces of data, we used the Euclidean distance to measure it. After the Hot-Deck Imputation, the R^2 score was 0.9993, which showed a significant improvement.

3.2.4. K-Means clustering imputation. As the K, which presents the number of clusters, needs to specified manually, we tried 22 numbers from 5 to 26. After finishing the imputation, the R^2 scores of the test set are showed as table 4.

Table 4. The R^2 scores of the test set with different numbers of the K-Means clusters.

Clustering Number	R^2 Score	Clustering Number	R^2 Score
5	0.9966	15	0.9975
6	0.9973	16	Nan
7	0.9966	17	0.9967
8	0.9972	18	0.9970
9	Nan	19	0.9986
10	0.9974	20	0.9988
11	0.9976	21	0.9970
12	Nan	22	Nan
13	0.9986	23	0.9990
14	Nan	24	0.9970

The R^2 scores fell to the range from 0.9966 to 0.9990, which means the K-Means clustering imputation worked not better.

3.2.5. Regression imputation based on the neural networks. Due to the small amount of data and the difficulty training a neural network with multiple outputs, we built a dual-hidden-layer feedforward network with single output. There are 100 neurons of each hidden layer, and the input layer possesses 7 neurons corresponding the factors except the factor to fill. 90% of the data was as training set, and the other 10% was as the test set. We got an ideal filling effect after 300 epochs as table 5 shows.

Table 5. The statistic of the regression imputation with the dual-hidden-layer feedforward network.

Item \ Missing Item	Reinforcement Volume Content	Interface Type	Interface Thickness	Porosity	Density
Training Set Count	266	294	236	92	115
Prediction Set Count	76	48	106	250	227

After the regression imputation with the dual-hidden-layer feedforward network above, we trained the prediction model with the CNN model above.

Finally, the R^2 score could be optimized to 0.9997. This result shows that the tensile property prediction model could get a significant improvement by the regression imputation based on a neural network.

As for the K-Means clustering imputation, we let the clustering number as 23, which leaded to the best R^2 score according to table 4, and compared the R^2 scores of different methods. The result is as table 6 shows.

Table 6. The comparison results of R^2 scores after different missing data process methods.

Zero Imputation (Ignore Missing Data)	Dropping Row	Mean Imputation	Hot Deck Imputation	K-Means clustering Imputation	Regression Imputation
0.9989	0.9880	0.9971	0.9993	0.9990	0.9997

Table 6 shows that the Hot-Deck Imputation and the regression imputation could get a better prediction model than ignoring the missing data.

4. Conclusion

This work collected the CMC experiments data from documents, built a CMC tensile property prediction model with missing data by building a dual-hidden-layer feedforward network and a 1-D convolution neural network. Then we experimented multiple methods to process incomplete data to research how to improve the robustness of the prediction model. By comparison, we found the Hot-Deck Imputation and the regression imputation based on a dual-hidden-layer feedforward network could improve the performance of the prediction model considerably. This work provides a valuable design reference for researching new CMCs.

We did not research deeper neural networks to build prediction model on account of the small amount of data. Further, we did not research the similarity, specificity and 3-D physical structure of different CMCs, which were only distinguished by encoding. In our opinions, it is a direction worthy of further study and improvement to mine and analyse these hidden factors.

Acknowledgments
Authors wishing to acknowledge assistance or encouragement from colleagues, special work by technical staff or financial support by the National Key R&D Program of China (2016YFB0700500) and the Science and Technology Project of Shaanxi Province (2018GY-048).

References

[1] Schmidt, S., et al., *Advanced ceramic matrix composite materials for current and future propulsion technology applications*. 2004. **55**(3-9): p. 409-420.
[2] Science, N. and T. Council, *Materials genome initiative for global competitiveness*. 2011: Executive Office of the President, National Science and Technology Council.
[3] Kim, J.K., W.A. Fuller, and W.R.J.T.A.o.A.S. Bell, *Variance estimation for nearest neighbor imputation for US census long form data*. 2011. **5**(2A): p. 824-842.
[4] MacQueen, J. *Some methods for classification and analysis of multivariate observations*. in *Proceedings of the fifth Berkeley symposium on mathematical statistics and probability*. 1967. Oakland, CA, USA.
[5] Troyanskaya, O., et al., *Missing value estimation methods for DNA microarrays*. 2001. **17**(6): p. 520-525.
[6] Enders, C.K., *Applied missing data analysis*. 2010: Guilford press.
[7] Naslain, R.R.J.C.P.A.A.S. and Manufacturing, *The design of the fibre-matrix interfacial zone in ceramic matrix composites*. 1998. **29**(9-10): p. 1145-1155.
[8] Beck, J.E. and B.P. Woolf. *High-level student modeling with machine learning*. in *International Conference on Intelligent Tutoring Systems*. 2000. Springer.
[9] Ioffe, S. and C.J.a.p.a. Szegedy, *Batch normalization: Accelerating deep network training by reducing internal covariate shift*. 2015.
[10] Suykens, J.A. and J.J.N.p.l. Vandewalle, *Least squares support vector machine classifiers*. 1999. **9**(3): p. 293-300.
[11] Rosenblatt, F., *Perceptions and the theory of brain mechanisms*. 1962: Spartan books.
[12] Goodfellow, I., Y. Bengio, and A. Courville, *Deep learning*. 2016: MIT press.
[13] Krizhevsky, A., I. Sutskever, and G.E. Hinton. *Imagenet classification with deep convolutional neural networks*. in *Advances in neural information processing systems*. 2012.
[14] Kingma, D.P. and J.J.a.p.a. Ba, *Adam: A method for stochastic optimization*. 2014.
[15] Glorot, X., A. Bordes, and Y. Bengio. *Deep sparse rectifier neural networks*. in *Proceedings of the fourteenth international conference on artificial intelligence and statistics*. 2011.
[16] Haykin, S.S., et al., *Neural networks and learning machines*. Vol. 3. 2009: Pearson education Upper Saddle River.
[17] Glantz, S.A. and B.K. Slinker, *Primer of applied regression and analysis of variance*. 1990.
[18] Draper, N.R. and H. Smith, *Applied regression analysis*. Vol. 326. 1998: John Wiley & Sons.
[19] Drucker, H., et al. *Support vector regression machines*. in *Advances in neural information processing systems*. 1997.

IOP Conf. Series: Materials Science and Engineering **647** (2019) 012005 doi:10.1088/1757-899X/647/1/012005

A Novel Coupler Design and Analysis with Shielding Material Tests for a CPT System of Electric Vehicles Based on Electromagnetic Resonant Coupling

J Duan[1,*] and W Wang[2]

[1,2] Department of Engineering and Design, University of Sussex, Brighton BN1 9QT, United Kingdom

* J.Duan@sussex.ac.uk.

Abstract. In this paper, a contactless power transfer (CPT) system using a novel geometrically enhanced energy transfer coupler with three different shielding materials has been built and analysed, along with the evaluations from aspects of electromagnetics and RMS power transmitting based on electromagnetic resonant coupling. A CPT system design improvement with the proposed H-shape ferromagnetic cores and the combined semi-enclosed passive electromagnetic shielding methods have been investigated in terms of generated electromagnetic field characteristics, system power transfer ratings, system efficiency optimization and performances of shielding materials. The results have shown that, across the range of operating frequency of the CPT system, aluminium shielding as a metallic material method could deliver better overall CPT system performance than other two ferromagnetic materials, steel 1010 and ferrite. In addition, the coupler prototype design limitations, misalignment tolerance and the passive shielding design considerations including distance between windings and inner surfaces of shielding shells have been discussed.

1. Introduction

Contactless power transfer (CPT) technologies based on inductive coupling have been developed to a series of applications in daily and industrial areas, such as small electronic devices, dynamic automotive battery systems, public transportations, medical implants, instrumentation test systems, and so on [1]-[5]. A conventional flat circular coils based contactless energy transmitting design was proposed for electric vehicles (EV) battery charging in [6], with introduction to inductively coupled technology. The compensation topologies have been evaluated in [7], which concluded that the series-series (SS) topology could perform the best. Among the proposed wireless energy transfer technologies, most of the studies regarded the CPT system issues as near-field transformer problems and analysed their models from equivalent circuits and electronics points of view. However, the CPT inductive coupling via air space or cores could be nonlinear scenarios due to permeability μ and B-H relations of the specific materials used within the coupling module, which causes that the mutual inductive coupling unit cannot be equivalent to a fixed electronic circuit like that of conventional ideal transformers.

In this paper, considering the changeable flux linkage and variable magnetic coupling against different operating frequencies, shielding materials and other factors in real CPT situations, the modelling and analysis based on electromagnetic field and 3D finite element methods (FEM) have been implemented in order to evaluate the three different shielding methods using ferromagnetic materials and conductive metallic materials. From aspects of both the electromagnetics and power electronics, the flux linkage, magnetic flux density, magnetic field strength and flux line distribution in real-time field have been investigated, along with power electronics indices which are induced currents, system output RMS real power, and system power transfer efficiency and so on. By comparisons of using steel 1010, ferrite and aluminium as shielding materials for the designed small-sized CPT system, aluminium shielding appears to yield the highest overall system perform within these three shielding methods.

2. CPT system design

2.1. System overview

An electromagnetic resonant coupling based CPT system consists of AC power supply, rectifier, high frequency inverter, capacitance compensation topology, magnetic coupler, high frequency rectifier, voltage regulator and load end. In order to optimize the RMS real power transfer and efficiency of the CPT system, the design of the magnetic coupling coil modular called coupler in this paper acts as a highly significant part in the overall system when the effectiveness of a constantly satisfactory magnetic flux generation and inductive coupling tend to occur at resonant conditions in between the space of coupler coils and ferromagnetic cores. In addition to the important role of the coupler, an efficient power electronic circuit including soft switching and active rectifier devices could also highly contribute to the CPT system performance [8]-[10].

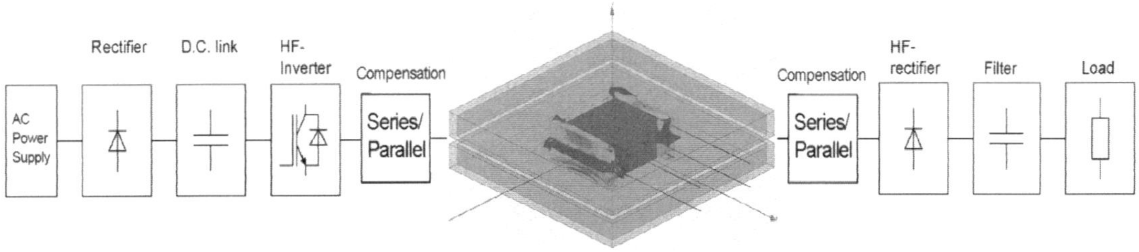

Figure 1. The H-shape coupler CPT system configuration with combined semi-enclosed shielding.

Within the space of inductive coupling, optimally shaping the flux line trajectories and forming the flux distributions using H-shape ferrite cores could expectedly improve the actual flux linkages and induced currents in real-world CPT scenarios. Practically, the high permeability and low core loss of ferrite materials could contribute to flux linkage and magnetic flux density in order for the CPT system to produce an optimized electromagnetic field performance especially when the electromagnetic resonant coupling tends to occur at specific conditions [11]. A stationary CPT system layout using the designed H-shape coupler with four compensation topologies (series-to-series, series-to-parallel, parallel-to-series and parallel-to-parallel) has been shown in figure 1. To achieve the magnetic resonance when the proposed CPT system tends to be stably operating, an appropriate capacitance compensation topology could be utilized to maximize energy transmitting capability between inductively coupled coils mounted on the designed H-shaped ferrite cores and as well as to minimize the magnitude of the real time values of reactance in the coupler. Thus the series-to-series (SS) compensation has been selected for the CPT system in this study, as shown in figure 2 below.

In order to analyse the inductive coupling performance of the proposed CPT system, based on RLC circuit theories, impedance matching principle and initial system tests with calculations for conditions of approaching electromagnetic resonant coupling, the designed system circuit parameters have been configured in figure 2, in which both the compensation capacitors are set with 150 nF, the load is set with a resistor of 50 ohm and a voltage source with an RMS value of 5 kV has been adopted as the system power supply.

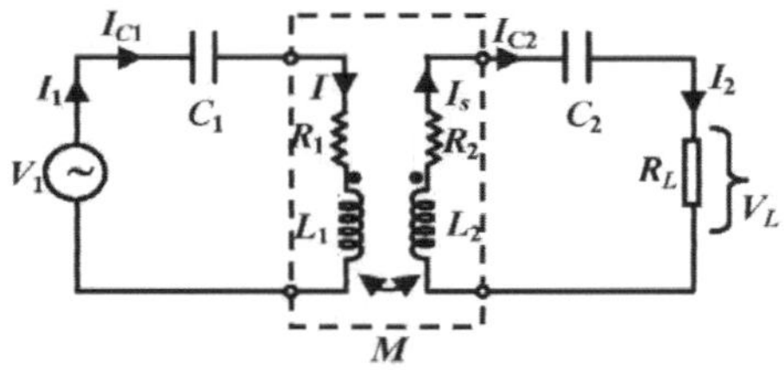

Figure 2. The simplified equivalent circuit of the CPT system with SS compensation topology.

2.2 *Geometric modeling for CPT coupler with semi-enclosed shielding*

Since the physical and geometric implementation of the coupling modular for the CPT system can definitively determine a maximum efficiency and highest transferred real power, the actual coupler design is important in the system configuration as well as considering the coupler size as one of the criteria.

To promote the magnetic flux density, flux linkage, mutual inductance and field strength, the specifically proposed H-shape core using Steel 1010 as ferromagnetic core material is designed to guide the magnetic flux along the expected paths. On the other hand, preventing energy loss in surrounding air [12] and conductive materials on the chassis of the vehicle and reducing the magnetic field flux leakage can be also contributive to the overall CPT system performance, which could be achieved by using passive magnetic shielding.

Due to the effects of high permeability and relative low eddy current loss of using ferromagnetic or conductive materials, the magnetic flux could be guided and formed by passive magnetic shielding methods [13]. Therefore, Steel1010, ferrite and aluminium have been utilized as the shielding materials in the Ansys 3D Maxwell platform based experiments in this paper. The Steel1010, as a ferromagnetic material, has nonlinear B-H curve. The ferrite adopted in this study has a relative permeability μ of 1000. The aluminium has a relative permeability μ of 1.000021.

In this paper, the H-shape coupler and the combined semi-enclosed passive shielding have been prototyped as presented in figure 3. The three different shielding materials are utilized to the CPT system coupler to output the performance results, respectively. Litz wire is selected for the windings in this model due to its skin effects reduction, proximity effects mitigation and capability for high frequency application. The small-sized magnetic coupler is designed as the geometric parameters given in table 1.

Table 1. The geometric characteristics of the H-shape coupler with passive shielding.

Parameters	Values
Winding size	100 mm x 100 mm x 20 mm
Core size	150 mm x 150 mm x 20 mm
H-shaped core bar size of each side	150 mm x 15 mm x 20 mm
Primary winding number of turns	80 turns
Secondary winding number of turns	80 turns
Air gaps of the CPT charging system	10 mm
Shielding size	400 mm x 400 x 30 mm
Shielding thickness	10 mm
Inner distance between shielding and coil	5 mm
Passive magnetic shielding materials	Steel 1010, Ferrite and Aluminium

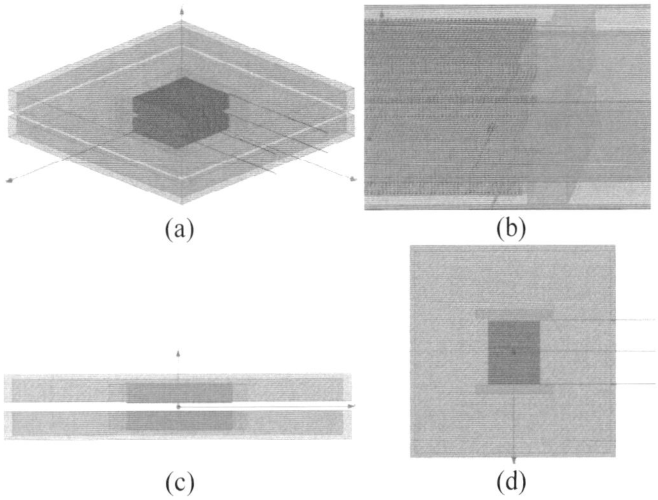

Figure 3. The H-shape coupler with the combined semi-enclosed shielding built in 3D FEM platform. (a) The CPT system coupler overview. (b) Detailed view of the windings and cores. (c) Left view of the system. (d) Top view of the system.

3. Basic principles and numerical representations of an electromagnetic resonant CPT system

To investigate and analyse the CPT system outputs with its expected electromagnetic coupling performance, the electric current density **J** induced on the secondary side of the coupler model is required to be determined. Hence, the Maxwell equations constitutively expressed by Ampere's law in equation (1), Gauss' law in equation (2), Faraday's law in equation (3) and the B-H curve relation in equation (4) are required to be solved:

$$\nabla \times \mathbf{H} = \mathbf{J} \tag{1}$$

$$\nabla \cdot \mathbf{B} = 0 \tag{2}$$

$$\nabla \times \mathbf{E} = -\frac{\partial \mathbf{B}}{\partial t} \tag{3}$$

$$\mathbf{B} = \mu \mathbf{H} \tag{4}$$

where **H** is the magnetic field strength in ampere per meter (A/m), **B** is the magnetic flux density in tesla (T), **E** is the electric field strength in volt per meter (V/m), and μ is the permeability non-linearly depending on local value of **B** in $\mathbf{B} = f(\mathbf{H})$. The electric current density **J** is in ampere per square meter (A/m^2).

$$\mathbf{B} = \nabla \times \mathbf{A} \tag{5}$$

Simultaneously with Ohm's law in electromagnetics by equation (6) and material equations of Pouillet's law by equation (7), the 3D finite-element method (FEM) is adopted to numerically solve Maxwell equations above and the relative material equations, using the methods of discretizing the modelled 3D space by tetrahedrons, translating the differential equations into algebraic equations [14]. The methods of semi-iterative conjugate gradient for each magnetic vector potential **A** [15], based on the initial B-H curve, constant values of μ can be selected for each finite element depending on local nonlinear value of **B** for the beginning. The flux density **B** values afterwards can be computed according to the calculated magnetic vector potential **A** by equation (5) derived from Gauss' law of equation (2). Repeatedly, the numerical computations can be completed until the convergence of the element permeability. It is known that

$$\mathbf{J} = \sigma \mathbf{E} \tag{6}$$

$$\mathbf{R} = \boldsymbol{\rho}\frac{\mathbf{l}}{\mathbf{C}} \tag{7}$$

where $\boldsymbol{\sigma}$ is the material-dependent parameter conductivity. $\boldsymbol{\rho}$ is the resistivity of the windings. **R** is known as electric resistance of the wire material. **l** is the total length of the wire. **C** is the cross-sectional area of the wire.

Based on the equations above, equation (1) turns into:

$$\nabla \times (\frac{1}{\mu}\nabla \times \mathbf{A}) = \mathbf{J} \tag{8}$$

$$\mathbf{E} = -\nabla\phi - \frac{\partial \mathbf{A}}{\partial t} \tag{9}$$

together with Faraday's law and Lorenz gauge vector potential by equation (9) [16], by using 3D FEM computations, the magnetic flux density **B**, the electric field strength **E** and the magnetic field with values of magnetic flux ϕ in weber can be numerically solved and determined.

In order to determine the self-inductance values of **L** in henry produced by the electromagnetic field on the both sides of the CPT system, the relations between the coil current values of i and the flux linkage λ satisfy:

$$\lambda_1 = N_1\phi_1 = L_1 i_1 \tag{10}$$

$$\lambda_2 = N_2\phi_2 = L_2 i_2 \tag{11}$$

$$L = \frac{\lambda}{i} \tag{12}$$

In addition, based on the RLC circuit theories [17] and circuit resonance conditions [18], the approximate natural resonant frequency for the designed coupling system in this paper can be determined in order to specify the power source operating frequencies, which then is for the analysis and optimization of the actual system performance.

To study the actual inductive magnetic coupling outcomes and electromagnetic field performance represented and reflected by the numerical vectors and scalars above, the actual RMS power generations given by equation (13) and the efficiencies given by equation (14) and (15) from the front end to the load end of the CPT system are required to be calculated, compared and analysed. cos φ is the power factor (PF) caused by phase difference φ between the induced voltage and current waveforms.

$$P_{RMS} = V_{RMS}\, I_{RMS} \left|\cos\varphi\right| \tag{13}$$

$$\eta_{Coupler} = \frac{P_{RMS_{Secondary\ coil}}}{P_{RMS_{Primary\ coil}}} \tag{14}$$

$$\eta_{Overal\ system} = \frac{P_{Load}}{P_{RMS_{Power\ supply}}} \tag{15}$$

4. Results and analysis

4.1. Flux linkage and currents with different shielding materials

In this investigation in terms of different ferromagnetic and conductive metallic shielding materials, the proposed H-shape coupler CPT system performance has been investigated across the major significant range of operating frequency and has been emphasized on the characteristic trends, from both the electromagnetics and electronics points of view. Theoretically, the flux linkage in electromagnetics and the induced current in power electronics have relatively reflective relations to the system performance based on Maxwell equations and induction principles.

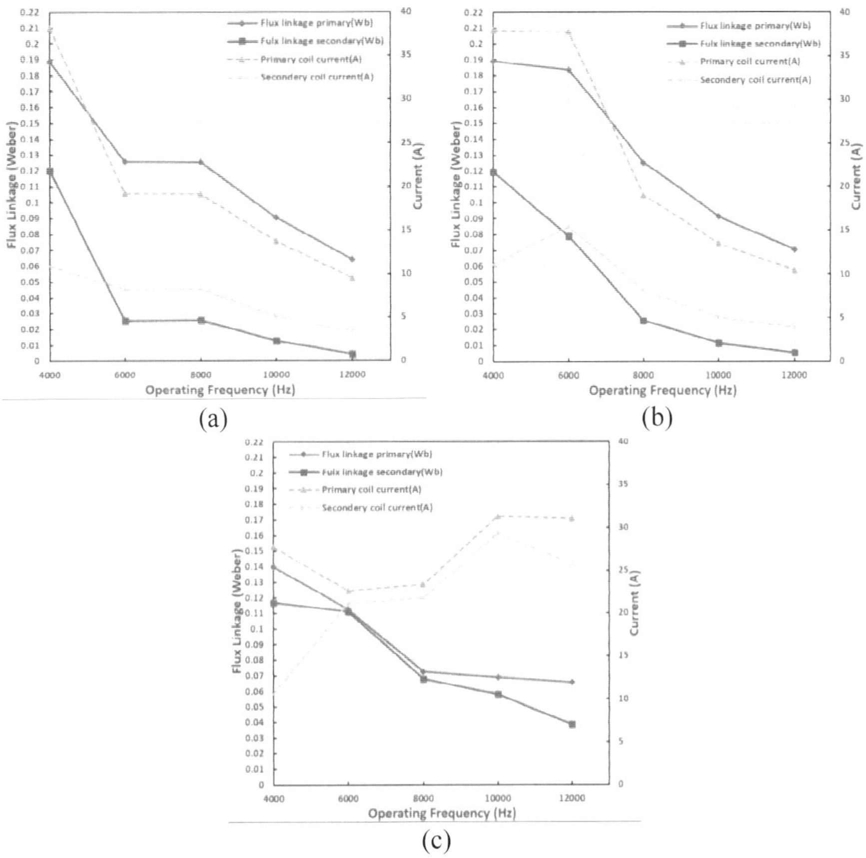

Figure 4. The flux linkage and current values of the designed CPT system versus system operating frequency. (a) Steel 1010 as shielding material. (b) Ferrite as shielding material. (c) Aluminium as shielding material.

From figures 4(a), (b) and (c), it can be seen that both the primary and secondary values of flux linkage for three shielding materials tend to decrease against the increase of system's operating frequency. Similarly, the primary and secondary coil currents for steel 1010 and ferrite in figure 4(a) and (b) show the decreasing trends. However, the values of primary and secondary coil currents tend to increase after 6000 Hz in figure 4(c), which may reflect better inductive capability of aluminium to produce satisfactory coupling for the system than steel 1010 and ferrite as passive shielding material.

In terms of the characteristic values, the flux linkage of primary and secondary coils for steel 1010 have shown relatively stable values of about 0.1257 Wb and 0.0258 Wb, respectively, corresponding to frequency range of 6000 to 8000 Hz in figure 4(a), after which these two values gradually tend to decrease. In the meantime, the induced currents of primary and secondary coils have the similar tendency like the corresponding flux linkages. It could be noticed that the values of primary and secondary currents tend to stabilize at about 19.13 A and 8.25 A, respectively, when the system operating frequency is supplied between 6000 Hz to 8000 Hz, in figure 4(a).

Regarding the flux linkage and current values for ferrite shielding in figure 4(b), both the values of the flux linkage of primary and secondary sides have a gradually moderating trend from 0.1890 Wb and 0.1195 Wb at 4000 Hz, respectively, down to 0.0703 Wb and 0.0054 Wb at 12000 Hz. While the primary coil current shows stable trend before 6000 Hz, after which a sharp drop can be seen against the operating frequency. However, the secondary coil current shows a peak value of 15.39 A at 6000 Hz, which is also the calculated optimal resonant coupling point for ferrite as shielding material.

With regard to shielding using conductive metallic material, aluminum shielding for the designed CPT system can be initially seen as a proper shielding method according to figure 4(c). The overall values of primary and secondary flux linkage are higher than steel1010 and ferrite despite the similar downward tendency. It could be seen that, after 8000 Hz, flux linkage values of primary and secondary sides show gradually decreasing change until 0.0581 Wb and 0.0384 Wb, respectively, at 12000 Hz, which is higher than the final and lowest values of steel 1010 and ferrite.

4.2. RMS real powers and efficiencies with different shielding materials

As indispensable indices for evaluating the CPT system performance, the RMS real power and efficiency across a range of system operating frequency could reflect the design effectiveness for the energy transmitting purpose of a system in real-world applications. In this paper regarding the three different shielding materials, the actual values of real transferred power and efficiency on both coil sides of the designed coupler are compared and analyzed.

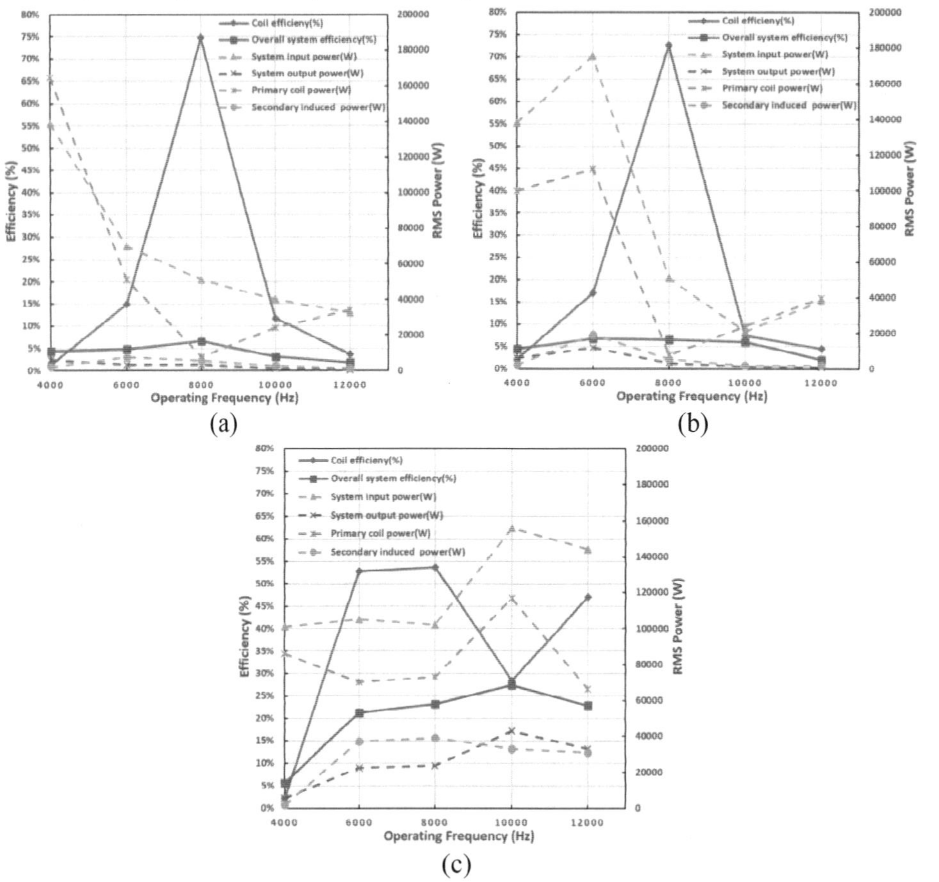

Figure 5. The RMS real powers and efficiencies of the designed CPT system versus system operating frequency. (a) Steel1010 as shielding material. (b) Ferrite as shielding material. (c) Aluminium as shielding material.

The actual electromagnetic field parameters generated are variable and the real-time phase angle differences are changeable in terms of various system operating frequencies [11], as well as the different shielding materials proposed in this paper, which determines the RMS real power and efficiency. Based on the actual power factor (PF) generated and original waveforms acquired after the CPT system outcomes stabilize at each testing operating frequency, the resulted values of RMS real power and efficiency can be worked out and analyzed.

From figure 5(a), it can be seen that the coil energy transfer efficiency could reach a peak value of 74.76% at 8000 Hz, at which point the system efficiency also reaches a maximum of 6.67%. The system input power significantly drops from 138.62 kW to 32.51 kW while the system output power on the load also reduces by about 10 times from 5.92 kW to 0.59 kW. It is noticeable that the primary coil power decreases to a lowest value of 7.51 kW at 8000 Hz whilst the secondary coil power could produce a second highest value on its curve, which causes the peak values 74.76% and 6.67%, respectively, for the coil efficiency and system efficiency.

Figure 5(b) depicts almost the same trend for the coil efficiency like in figure 5(a), whilst the overall system efficiency show a stably higher range of values over 6% across the operating frequency range between 6000 Hz to 10000 Hz. The overall system efficiency reaches its maximum of 6.77% at 6000 Hz. Importantly, it can be noticeably seen that the system input power, system output power on load, primary coil power and secondary coil achieve their maximum values of 175.19 kW, 11.85 kW, 112.11 kW and 19.11 kW, respectively, at the calculated magnetic resonant coupling frequency of about 6000 Hz.

Regarding the performance of aluminium shielding, it can be noticed that almost all the results of the system could show improvements in figure 5(c). Significantly, the overall system efficiency could reach a peak value of 27.48% at 10000 Hz and could also stay at stably satisfactory values of over 20% when the operating frequency is supplied with 6000 Hz and above. The system output power on load end achieves a maximum of 42.90 kW at 10000 Hz.

Therefore, the aluminium shielding performs with the highest results of criteria within the three shielding materials, which indicates satisfactory power transfer rating, efficiency in addition to acceptable electromagnetic field flux distributions and coupling performance.

4.3. Analysis and comparison on main results with different shielding materials based on electric vehicle charging

In addition to graphic trend analysis for each model with different shielding methods above, table 2 below could indicate the major critical results for comparing and evaluating the CPT system outcomes and performance. As known, the load current could firstly reflect the effectiveness of inductive coupling phenomena, by which it can been seen that the CPT system with aluminium could produce highly effective inductive coupling for electric vehicle contactless charging applications than systems with steel 1010 and ferrite shielding according to numerical comparisons in table 2.

Generally, electric vehicles (EV) could gain faster charging depending on higher CPT system output real power rating. From table 2, the system output power of steel1010 shielding is unable to deliver a rating value of over 1 kW, while the ferrite shielding CPT system can produce 11.85-kW system output power on load at its resonant coupling status with the calculated natural resonant frequency of about 6000 Hz. Noticeably, the CPT system with aluminium shielding could generate relatively high system output real power across a wider range of operating frequency, which contributes to a maximum value of 42.90 kW when the CPT system performs to stabilize, at the resonant coupling frequency of about 10000 Hz. When considering the energy use efficiency from the power supply, the overall system efficiency could be of reference. By comparison, the aluminium shielding CPT system is able to produce relatively high system efficiency especially at 10000 Hz as presented in table 2.

Table 2. Main numerical results of the H-shape coupler CPT system with threes passive shielding materials.

Operating Frequency (Hz)	CPT system with Steel1010 Shielding			CPT system with Ferrite Shielding			CPT system with Aluminium Shielding		
	Load Current (A)	System Output Power (W)	Overall System Efficiency	Load Current (A)	System Output Power(W)	Overall System Efficiency	Load Current (A)	System Output Power (W)	Overall System Efficiency
4000	10.8830	5921.6438	4.2718%	11.0816	6140.1916	4.4533%	10.6396	5659.8033	5.6023%
6000	8.2543	3406.8649	4.8806%	15.3962	11852.4135	6.7656%	21.1614	22389.6902	21.3399%
8000	8.2598	3411.4717	6.6690%	8.1282	3303.6662	6.4895%	21.7739	23704.4088	23.2055%
10000	5.0780	1244.7123	3.0932%	5.0251	1262.4810	6.0493%	29.2918	42899.5869	27.4804%
12000	3.4414	593.2719	1.8249%	3.8790	752.3685	1.9919%	25.8064	33008.5921	22.9075%

4.4. Analysis on a selected CPT system model with aluminum shielding method based on electromagnetic field

Based on comparisons and analysis in previous sections, it can be found that the H-shape coupler CPT system with aluminium shielding could provide highest power electronic performance in the three shielding methods. Thus, the CPT system with aluminium shielding, at its optimal performance frequency of 10000 Hz, has been selected to be analysed from aspects of the electromagnetics.

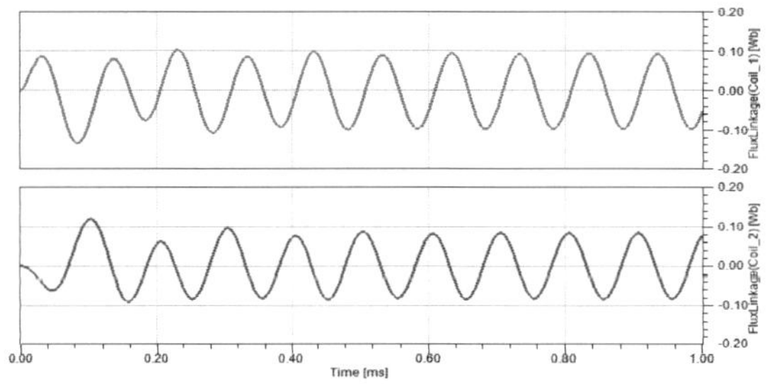

Figure 6. The flux linkage waveforms of the designed H-shape coupler CPT system with aluminium passive shielding, at the operating frequency of 10000 Hz.

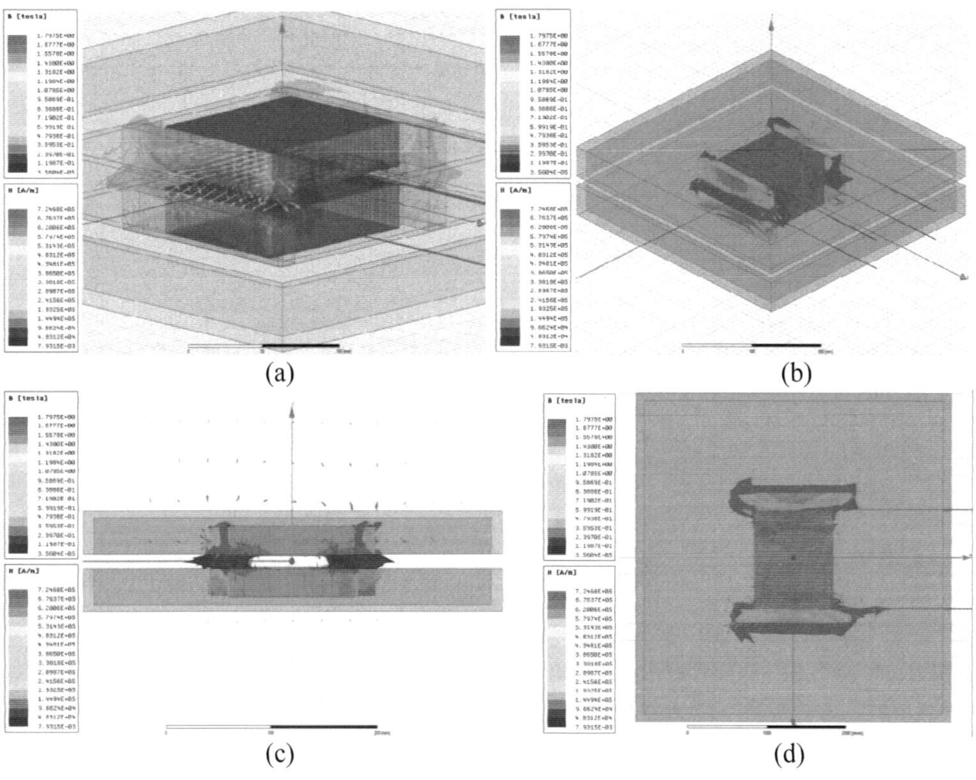

Figure 7. The electromagnetic field of the designed H-shape coupler CPT system with aluminium passive shielding at 1 *ms*, indicating the magnetic field strength **H** and the magnetic flux density **B**, at 10000 Hz. (a) The **B** vector in the field. (b) The dimitric view of the field. (c) The left view of the field. (d) The top view of the field.

As can be seen from figure 6, the alternating flux linkage waveforms of the primary and secondary coils are able to be sinusoidally stable since the sixth period, which means the CPT system could tend to perform immediately with electromagnetic coupling stability within 1 *ms* after the system is activated to be magnetically coupled. Once the electromagnetic stability is established, the RMS values of the primary and secondary flux linkage could be 0.0689 Wb and 0.0581 Wb, corresponding to the measured amplitude values of 0.0974 Wb and 0.0822 Wb, respectively. Besides, it also could be computed that the phase difference between the two AC flux linkages tends to be about 108 degrees when the system reaches an AC stable and magnetic resonant coupling.

In terms of the electromagnetic field parameters and performance, the CPT system with aluminium shielding at the operating frequency of 10000 Hz has been selected to be investigated as a typical magnetic resonant coupling scenario. As shown in figure 7(a), the vector arrows of the magnetic flux density **B** show the expected directions, mostly concentrating at the external edges and corners of the designed H-shaped core, along with maximum scalar magnitude of 1.7975 Teslas, In the meantime, the magnetic field strength **H** reaches its peak value of 7.2468 x 10^5 A/m at the ending point of the external corners of the core, which also indicates the effectiveness of the geometric H-shape design for shaping and guiding the flux lines.

According to figures 7(b), (c) and (d), the vector field **B** and **H** are established to be alternatively varying within the combined semi-enclosed air space by the effects of aluminium passive magnetic shielding, which can prove that the shielding does prevent the magnetic field flux distributions from spreading and leaking to the remote surrounding space. In real-world applications, this effect may consequently reduce energy losses to the air, to the surrounding conductive elements of electric vehicle chassis or even to the human driver as a potential EMI issue when in a high frequency range.

5. Discussion

As demonstrated and illustrated in the electromagnetic field of the designed H-shape coupler CPT system, the ferromagnetic material as the core with high permeability and H-shape can produce and form the magnetic flux distribution as expected, especially facilitating the magnetic flux density and magnitude field strength at the external corners and edges of the H-shape core. The optimally guided and formed flux vector trajectories and field distribution can be utilized and has been proved to promote the CPT system coupling performance from perspectives of electromagnetics.

By comparing the obtained power transfer ratings and system efficiency values of three shielding methods using different ferromagnetic and conductive materials, it can be found that deploying aluminium as shielding material could output relatively more satisfactory CPT system performance. The H-shape coupler CPT system using aluminium shielding enables to deliver a maximum overall system output RMS real power of 42.90 kW and to produce a peak system efficiency of 27.48% at about the natural resonant frequency of 10000 Hz. Besides, at higher range of operating frequencies, the CPT system with aluminium shielding appears to generate sustainable satisfactory system efficiency, transfer power rating on load and flux linkage, which could be of advantage for real-world electric vehicle contactless charging applications.

With regard to limitations, the distance between the windings and inner shielding surfaces in this prototype may be too small, which would probably have caused the ultimately optimal magnetic field flux distribution and flux linkage values to be restricted to be achieved, to some degree. In the further investigations, the distance between the windings and inner shielding surfaces is supposed to be enlarged from 5 mm to 15 mm under the preconditions of expected shielding functions and limited space usage, to find out the optimized inner shielding distance for the CPT system.

In real-world applications, the actual electromagnetic field performance and inductive coupling quality can be significantly affected by how the electric vehicle is parked over the ground side power transmitting module with longitudinal and lateral misalignments, which should be counted as system tolerance to parking misalignment. In addition, the module size of CPT system will have to be optimized to be completely suitable to real-world vehicle chassis for installation and application.

6. Conclusion

In this paper, a small-sized geometrically improved H-shape coupler CPT system and the basic principles of electromagnetic resonant coupling model have been presented, along with the analysis and discussion on RMS power transfers, CPT system efficiencies, generated electromagnetic field characteristics and performances of using three passive shielding materials. It can be found that the magnetic field flux distributions have been optimally formed by the designed ferromagnetic H-shape core. By using aluminium passive shielding to facilitate the field flux paths, the CPT system could output a maximum overall system efficiency of 27.48% and a peak RMS real power of 42.90 kW on the system load end, at about 10000-Hz operating frequency when the CPT system tends to perform with magnetic resonant coupling status. In addition, the limitations of the designed CPT system and some future focuses have been discussed.

References

[1] Budhia, M., Vyatkin, V. & Covic, G.A., 2008. Powering flexible manufacturing systems with intelligent contact-less power transfer. *2008 6th IEEE International Conference on Industrial Informatics*, pp.1160–1165.

[2] Pengfei Li, R. & Bashirullah, 2007. A Wireless Power Interface for Rechargeable Battery Operated Medical Implants. *IEEE Transactions on Circuits and Systems II: Express Briefs*, 54(10), pp.912–916.

[3] Radecki, A. et al., 2011. 6W/25mm2 inductive power transfer for non-contact wafer-level testing. *2011 IEEE International Solid-State Circuits Conference*, pp.230–232.

[4] Kurs, A. et al., 2007. Wireless power transfer via strongly coupled magnetic resonances. *Science (New York, N.Y.)*, 317(5834), pp.83–86.

[5] Hui, S.Y.R. & Ho, W.W.C., 2005. A new generation of universal contactless Battery Charging platform for portable Consumer Electronic equipment. *IEEE Transactions on Power Electronics*, 20(3), pp.620–627.

[6] Chwei-Sen Wang, O., Stielau & Covic, 2005. Design considerations for a contactless electric vehicle battery charger. *IEEE Transactions on Industrial Electronics*, 52(5), pp.1308–1314.

[7] Villa, J.L. et al., 2012. High-Misalignment Tolerant Compensation Topology for ICPT Systems. *IEEE Transactions on Industrial Electronics*, 59(2), pp.945–951.

[8] Hosotani, T. & Awai, I., 2012. A novel analysis of ZVS wireless power transfer system using coupled resonators. *2012 IEEE MTT-S International Microwave Workshop Series on Innovative Wireless Power Transmission: Technologies, Systems, and Applications*, pp.235–238.

[9] Yang Sun et al., 2011. A high speed comparator based active rectifier for wireless power transfer systems. *2011 IEEE MTT-S International Microwave Workshop Series on Intelligent Radio for Future Personal Terminals*, pp.1–2.

[10] Kim, S. et al., 2010. Semi-active high-efficient CMOS rectifier for wireless power transmission. *2010 IEEE Radio Frequency Integrated Circuits Symposium*, pp.97–100.

[11] Duan, J. & Wang, W., 2018. Electromagnetic Coupling Optimization by Coil Design Improvements for Contactless Power Transfer of Electric Vehicles. In Proceedings of the Future Technologies Conference (FTC) 2018. *Advances in Intelligent Systems and Computing*, vol 881, pp. 944-958.

[12] Paul, C.R., 1992. *Introduction to electromagnetic compatibility*, New York: Wiley.

[13] Wohlfarth, E. P., 1982. *Ferromagnetic Materials: A Handbook on the Properties of Magnetically Ordered Substances*. Amsterdam: North-Holland.

[14] Sadiku, M., 1992. *Numerical techniques in electromagnetics*, Boca Raton: CRC P.

[15] Silvester, P., Ferrari, R., 1990. *Finite elements for electrical engineers* 2nd ed., C.U.P.

[16] Jackson, J.D. & Okun, L.B., 2000. Historical roots of gauge invariance. *Reviews of Modern Physics*, 73(3), pp.663–680.

[17] Grover, F., 1962. *Inductance Calculations: Working Formulas and Tables*. Dover.

[18] Rashid, M.H., 2004. *Power Electronics: circuits, devices, and applications* 3rd ed., Upper Saddle River, N.J.: Prentice Hall.

IOP Conf. Series: Materials Science and Engineering **647** (2019) 012006 doi:10.1088/1757-899X/647/1/012006

Modeling the Electrical Conductivity of $Ni_{1-x}Fe_x$-SDC Composite Anode by Using PSO-SVR

J. L Tang and R. F Yang*

Department of Physics and Engineering Technology, Guilin Normal College, Guilin, Guangxi, 541199, P.R. China

* Email: tantian62@hotmail.com

Abstract. Studies have shown that numerous indexes affecting the electrical conductivity of Solid Oxide Fuel Cell (SOFC) anode. In order to improve performance of SOFC, it is advantageous to have a model with which one can modeling the electrical conductivity at different operating conditions. In this study, a model utilizing support vector regression (SVR) approach combined with particle swarm optimization (PSO), was proposed to modeling the electrical conductivity of $Ni_{1-x}Fe_x(0 \leqslant x \leqslant 0.25)$-$Ce_{0.85}Sm_{0.15}O_{2-\delta}$(SDC) composite anode. The test result by PSO-SVR show that the root mean square error (*RMSE*) of test samples is 3.79, mean absolute percentage error (*MAPE*) of test samples is 0.82%, multiplecorrelation correlation coefficients (R^2) of test samples is 1.00, which is satisfied with the engineering demand. The result of this investigation provides that PSO-SVR is an effective tool for modeling the electrical conductivity of $Ni_{1-x}Fe_x$-SDC composite anode.

1. Introduction

Solid Oxide Fuel Cell (SOFC) is a class of fuel cells. It can produces electricity directly from oxidizing a fuel. Advantages of this class of fuel cells include high efficiency, low emissions, completely solid component, fuel adaptability, and relatively low cost [1]. Studies have shown that numerous indexes affecting the electrical conductivity of SOFC anode. In order to improve performance of SOFC, it is advantageous to have a model with which one can modeling the electrical conductivity of SOFC anode at different operating conditions.

As a supervised learning method, Support Vector Regression (SVR) proposed by Vapnik and co-workers. Unlike most of the traditional methods, SVR is a new method of soft sensor modelling based on Statistical Learning Theory (SLT) [2]. Research shows that SVR with many remarkable characteristics, such as fast-learning, resistant to the over-fitting problem, and excellent generalization performance for the small-sample dataset. In the last few years, SVR has been successfully applied to solve modeling problems in fuel cell fields [3, 4].

In this study, the SVR model was proposed to modeling the electrical conductivity of $Ni_{1-x}Fe_x(0 \leqslant x \leqslant 0.25)$-$Ce_{0.85}Sm_{0.15}O_{2-\delta}$(SDC) composite anode according to the SOFC electrical conductivity dataset which was measured under different operating temperature and Fe content in $Ni_{1-x}Fe_x$-SDC composite anode by J. P Niu. Particle Swarm Optimization (PSO) is used in this study for searching the best parameter set of SVR for improving the accuracy of the SVR model.

2. Material and methods

2.1. Description of the SOFC
The typical structure of a single SOFC is consists of anode, electrolyte membrane and cathode. It is shown in Figure 1[5].

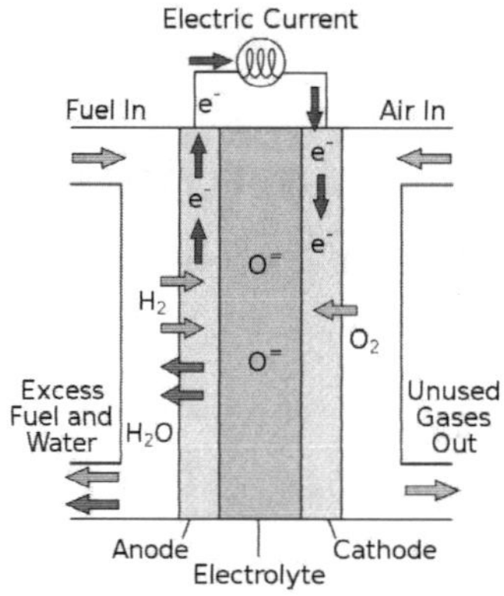

Figure1. Schematic of an Individual SOFC

The electrochemical reactions at the anode in a single SOFC are below:

Anode reactions: $2O^{2-} + 2H_2 - 4e^- \rightarrow 2H_2O$

OR: $4O^{2-} + CH_4 - 8e^- \rightarrow 2H_2O + CO_2$

$O^{2-} + CO - 2e^- \rightarrow CO_2$

2.2. SVR theory [6]
For SVR, the basic idea is to map X from the input space into a higher-dimensional feature space F via a nonlinear mapping function $\Phi(x)$, and then to conduct linear regression in F space. Therefore, SVR is to find the linear relation equation (1) based on a given dataset $(x_1, y_1), \ldots, (x_n, y_n)$.

$$f(\boldsymbol{x}) = \boldsymbol{w} \cdot \Phi(\boldsymbol{x}) + b,\ \Phi: \boldsymbol{R}^n \rightarrow \boldsymbol{F},\ \boldsymbol{w} \in \boldsymbol{F}. \quad (1)$$

where w is a vector for regression coefficients, b is a bias. They are estimated by minimizing the regularized risk function $R(C)$, namely:

$$\text{minimize } R(C) = 1/2\|\boldsymbol{w}\|^2 + C\sum_{i=1}^{n} L_{\varepsilon}(f(\boldsymbol{x}_i) - y_i), \quad (2)$$

$$L_{\varepsilon}(f(\boldsymbol{x}_i) - y_i) = \begin{cases} 0, & \text{if } |f(\boldsymbol{x}_i) - y_i| < \varepsilon, \\ |f(\boldsymbol{x}_i) - y_i| - \varepsilon, & \text{if } |f(\boldsymbol{x}_i) - y_i| \geq \varepsilon. \end{cases} \quad (3)$$

where C is a regularized factor, n is the number of training samples, ε is a prescribed parameter controlling the tolerance to error. After solved the regression function (1) has the following explicit form:

$$f(\boldsymbol{x}) = \sum_{i=1}^{l} (\alpha_i - \alpha_i^*) k(\boldsymbol{x}, \boldsymbol{x}_i) + b, \quad (4)$$

In equation (4), $k(\boldsymbol{x}, \boldsymbol{x}_i) = \Phi(\boldsymbol{x}) \cdot \Phi(\boldsymbol{x}_i)$ is a kernel function, α_i and α_i^* are Lagrange multipliers. Choosing different kernel function can generate different SVR models. In this paper, the radial basis kernel (5) was utilized:

$$k(\boldsymbol{x}, \boldsymbol{x}_i) = \exp(-\gamma \|\boldsymbol{x} - \boldsymbol{x}_i\|^2), \quad (5)$$

2.3. Choosing of SVR parameters by PSO

PSO is an evolutionary computing technology designed by Kennedy and Eberhart in 1995[7]. It has been applied successfully to various optimization problems.

In this study, PSO was introduced to search the optimal subset (ε, C, γ) of SVR [8]. PSO searches the best parameter subset (ε, C, γ) of SVR by regulating velocity and location of particles. Each particle is made up of a parameter vector (ε, C, γ). The ith particle is looked as a point in the 3D space and represented as $\boldsymbol{u}_i = (u_{i1}, u_{i2}, u_{i3})^T$, its velocity is represented as $\boldsymbol{v}_i = (v_{i1}, v_{i2}, v_{i3})^T$, the position of each particle with its best-fit value that is its local best, is remembered and denoted as $\boldsymbol{p}_{ibest}$, its global best, which is the position with the best-fit value of all particles, is also recorded as $\boldsymbol{g}_{best}$. At each iterative process, the velocity and position of each particle was adjusted by tracking its local best value, global best value and its present velocity, their iterative equations are as follows:

$$\begin{aligned} \boldsymbol{v}_i(t+1) = \omega \cdot \boldsymbol{v}_i(t) + c_1 \cdot \mathrm{rand}() \cdot (\boldsymbol{p}_{ibest} - \boldsymbol{u}_i(t)) + \\ c_2 \cdot \mathrm{rand}() \cdot (\boldsymbol{g}_{best} - \boldsymbol{u}_i(t)) \end{aligned} \quad (6)$$

$$\boldsymbol{u}_i(t+1) = \boldsymbol{u}_i(t) + \boldsymbol{v}_i(t+1) \quad (7)$$

where $\boldsymbol{v}(t)$, $\boldsymbol{v}(t+1)$, $\boldsymbol{u}(t)$, $\boldsymbol{u}(t+1)$ are respectively the speed and position of present moment and the next moment; rand() is a random value between 0 and 1; c_1 and c_2 are both learning factors; ω is a weighting factor to accelerate the convergence rate, its value should be automatically regulated with the iterative time of algorithm extending, defined generally as:

$$\omega = \omega_{min} + (iter_{max} - iter) \cdot (\omega_{max} - \omega_{min}) / iter_{max} \quad (8)$$

where ω_{max} and ω_{min} are the biggest and smallest weighting factors respectively, $iter$ is the number of current iteration. $iter_{max}$ is the total number of iterations.

2.4. Dataset

The dataset used in this study was generated by J. P Niu [9] and is tabulated in Table 1. This dataset includes the electrical conductivity of $Ni_{1-x}Fe_x$-SDC composite anode for 42 samples in different operating temperatures (T (°C)) and Fe contents (x = 0, 0.05, 0.10, 0.15, 0.20, 0.25). The atomic mole ratio of Ni and Fe is 1:0, 0.95:0.05.0.90:0.10, 0.85:0.15, 0.80:0.20, 0.75:0.25, labeled F0, F1, F2, F3, F4,F5.

Table 1 The electrical conductivity of $Ni_{1-x}Fe_x$-SDC composite anode at different operating temperatures.

T (°C)	Sample					
	F0 δ(S· cm^{-1})	F1 δ(S· cm^{-1})	F2 δ(S· cm^{-1})	F3 δ(S· cm^{-1})	F4 δ(S· cm^{-1})	F5 δ(S· cm^{-1})
500	1448.579	1770.277	1057.153	810.173	858.632	639.792
550	1384.018	1705.553	1022.898	713.234	752.388	569.120
600	1306.730	1644.067	995.233	680.070	679.976	505.219
650	1268.693	1587.272	959.262	659.398	657.491	478.010
700	1213.537	1527.940	930.118	642.612	642.972	468.367
750	1170.925	1485.234	904.463	628.380	631.479	462.814
800	1142.038	1452.910	884.672	617.844	621.775	462.724

2.5. Modeling and results

In the SVR model, the operating temperature and Fe content were employed as input variables, while as the electrical conductivity of $Ni_{1-x}Fe_x$-SDC composite anode as output variable.

Forty samples were selected as training samples, the other two samples numbered acted as the test samples.

Table 2. The modeling results by PSO-SVR.

NO	Input		Output		
	operating temperature (°C)	Fe contents	electrical conductivity (S· cm^{-1})	modeling results (S· cm^{-1})	percentage error (%)
1	500	0.00	1448.579	1448.5790	0
2	550	0.00	1384.018	1384.0190	0
3	600	0.00	1306.730	1306.7300	0
4	650	0.00	1268.693	1268.6940	0
5	700	0.00	1213.537	1213.5370	0
6	750	0.00	1170.925	1170.9250	0
7	800	0.00	1142.038	1142.0390	0
8	500	0.05	1770.277	1770.2760	0
9	550	0.05	1705.553	1705.5530	0
10	600	0.05	1644.067	1644.0670	0
11	650	0.05	1587.272	1587.2720	0
12	700	0.05	1527.940	1527.9390	0
13	750	0.05	1485.234	1485.2340	0
14	800	0.05	1452.910	1452.9100	0
15	500	0.10	1057.153	1057.1540	0
16	550	0.10	1022.898	1022.8990	0
17	600	0.10	995.233	995.2342	0
18	650	0.10	959.262	959.2632	0
19	700	0.10	930.118	930.1190	0
20	750	0.10	904.463	904.4643	0
21	800	0.10	884.672	884.6732	0
22	500	0.15	810.173	810.1731	0
23	550	0.15	713.234	713.2339	0
24	600	0.15	680.070	680.0700	0
25	650	0.15	659.398	659.3980	0
26	700	0.15	642.612	642.6120	0
27	750	0.15	628.380	628.3803	0
28	800	0.15	617.844	617.8439	0
29	500	0.20	858.632	858.6325	0
30	550	0.20	752.388	752.3883	0
31	600	0.20	679.976	679.9764	0
32	650	0.20	657.491	657.4918	0
33	700	0.20	642.972	642.9728	0
34	750	0.20	631.479	631.4793	0
35	800	0.20	621.775	621.7758	0
36	500	0.25	639.792	639.7921	0
37	550	0.25	569.120	569.1200	0
38	600	0.25	505.219	505.2190	0
39	650	0.25	478.010	478.0104	0
40	700	0.25	468.367	468.3676	0
41*	750	0.25	462.814	466.6071	0.82
42*	800	0.25	462.724	458.9360	-0.82

* Test sample

2.6. Evaluation of model's performance

Three indices, root mean square error (*RMSE*), mean absolute percentage error (*MAPE*) and multiplecorrelation coefficients (R^2) were adopted for performance evaluation. They are formulated by equations. (9), (10) and (11) respectively:

$$RMSE = \sqrt{\frac{1}{m}\sum_{i=1}^{m}(\hat{y}_i - y_i)^2} \tag{9}$$

$$MAPE = \frac{1}{m}\sum_{j=1}^{m}\left|\frac{\hat{y}_j - y_j}{y_j}\right| \tag{10}$$

$$R^2 = \left[\sum_{j=1}^{m}(y_j - \bar{y})(\hat{y}_j - \bar{\hat{y}})\right]^2 \Big/ \sum_{j=1}^{m}(y_j - \bar{y})^2 \cdot \sum_{j=1}^{m}(\hat{y}_j - \bar{\hat{y}})^2 \tag{11}$$

Where m denotes the number of samples, y_j represents the jth target value, $\hat{y}_j$ stands for the predicted value for the jth test sample, $\bar{\hat{y}}$ is the mean value of the predicted values $\hat{y}_j$ (j=1~ m) for samples.

Table 3. Performance of PSO-SVR model.

	RMSE	*MAPE* (%)	R^2
Training samples	0.00	0.00	1.00
Test sample	3.79	0.82	1.00

2.7. Analysis and discussions

In this study, the optimal parameter subset (ε, C, γ) of PSO-SVR model were (0.000002, 4238671.400509, 94.790245).

From Table 2, it can be observed that, all the percentage error for the 40 training sample's electrical conductivity of $Ni_{1-x}Fe_x$-SDC composite anode is 0. The percentage error for the 2 test sample's electrical conductivity of $Ni_{1-x}Fe_x$-SDC composite anode is no more than $\pm 1\%$.

Table 3 reveals that the *RMSE* of 40 training samples is 0, the *MAPE* is 0, R^2 as high as 1.00. The *RMSE* of 2 test samples comes up to 3.79, the *MAPE* is 0.82%, R^2 reach 1.00 too.

All these results indicate, the performance of PSO-SVR is excellent, it enough to meet the engineering demand.

3. Conclusions

In this study, the PSO-SVR model was established to modeling the electrical conductivity of $Ni_{1-x}Fe_x$-SDC composite anode under two influence factors, including operating temperature (*T*) and Fe content in $Ni_{1-x}Fe_x$-SDC composites anode. The result is revealed that: the generalization ability of PSO-SVR model is high enough. The PSO-SVR is a promising and practical methodology to modeling the electrical conductivity of $Ni_{1-x}Fe_x$-SDC composite anode.

Acknowledgments

This work is supported by the 2018 guangxi university young and middle-aged teacher basic ability enhancement project (Application of support vector regression technique in concrete strength prediction (2018KY0916))

References

[1] Badwal, S P S 2014 Review of Progress in High Temperature Solid Oxide Fuel Cells *J Aust Ceram Soc.* 50 (1) pp 23-37.
[2] Vapnik V N 1995 *The nature of statistical learning theory*. (New York: Springer) .
[3] Tang J L, Cai C Z, Xiao T T and Huang S J 2012 Modeling and predicting the composite of composite cathode for solid oxide fuel cell by using support vector regression support vector regression *Int J Mod Phys B* 26 p1250093 (9 pages)

IOP Conf. Series: Materials Science and Engineering **647** (2019) 012006 doi:10.1088/1757-899X/647/1/012006

[4] Tang J L and Huang J 2015 Modeling of Proton Exchange Membrane fuel cell using support vector regression machine *5th International Conference on Information Engineering for Mechanics and Materials* pp380–5.
[5] http://tech.newmaker.com/art_23205.html.
[6] Wen Y F, Cai C Z, Liu X H, Pei J F, Zhu X J and Xiao T T 2009 Corrosion rate prediction of 3C steel under different seawater environment by using support vector regression *Corros Sci.* 51 pp. 349-55.
[7] Kennedy J, Eberhart R, Particle swarm optimization, J. Proceedings of IEEE International Conference On neural Networks, Perth, Australia, (1995), 1942-1948.
[8] Tang J L, Liu H Y, Gui M H and Tang J Y 2017 SVR-Based Predictive Model for 2519 Aluminum Alloy *Key Eng Mater*. 719 pp 41-5.
[9] Niu J P 2008 The preparation and performance study of Solid Oxide Fuel Cells composite anode materials D (china: Jilin University)

IOP Conf. Series: Materials Science and Engineering **647** (2019) 012007 doi:10.1088/1757-899X/647/1/012007

Mechanism and Research on Preparation of AlN Powder by Carbon Thermal Reduction Method

L Y Wang[1] and N Zhang[2]

[1]Architecture & Civil Engineering Institute, Shenyang University, Liaoning, 110000, China
[2]Liaoning key laboratory of advanced materials

Corresponding author: E-mail address: w.ly428@163.com

Abstract. The alumina and activated carbon used as raw materials were prepared by chemical precipitation method. The uniformed aluminum nitride precursor was synthesized by mechanical mixing process. The effects of nitride temperature and time on the diameter of the particle were studied in detail. Ultrafined aluminum nitride powder was prepared by carbon-thermal reduction method. The results show that the precursor has better activity and nitride reaction proceeds rapidly. The complete conversion can be got at 1600°C for 2h. After calcination, it can be seen from the XRD graph that the powder with uniform size distribution has high AlN content.

1.Introduction

Aluminum nitride with high temperature corrosion resistance and temperature stability, higher strength and hardness has potentialities on the application of high temperature structural materials.

At present, with the development of China's national economy, the demand for aluminum nitride shows an increasing trend year by year, so the preparation of ultra-fine aluminum nitride powder has very important practical significance [1].

In this paper, AlN powder is prepared by carbon thermal reduction method to solve the technical problems such as insufficient research on ultra-fine aluminum nitride produced in mass production, unsuitable technological process for industrial producation and poor dispersibility of products.

2. Reaction mechanism for synthesis of AlN powder by carbon thermal reduction

The AlN powder prepared by carbothermal reduction has higher purity, smaller size and active soybean. For its inexpensive starting material and stable process, the carbothermal reduction method becomes an aborbed synthesis method and attracts people's widespread attention. However, due to the poor binding property between alumina and carbon, it is difficult to achieve uniform mixing. In addition, the low reactivity of alumina itself requires high reaction temperature, which increases the production cost, which restricts the promotion and application of AlN powder.

Improving the bonding property of aluminum and carbon and exploring the reaction mechanism of the system are the key points of carbon thermal reduction. As intermediate phases such as Al(g), $Al_2O(g)$ and $Al_2O_3(g)$ will be produced in Al_2O_3-C-N_2 system, other substances other than AlN phase, such as AlON, may also appear in the reaction products, so the research on reaction mechanism of preparing AlN by carbon thermal reduction becomes complicated and there are many differences. In general, the reaction mechanism can be divided into gas-solid reaction and solid-solid reaction according to the types of the first reaction in the process of carbon thermal reduction [2].

Different hypotheses were exist for the specific reaction process in the Al_2O_3-C-N_2 reaction system.

For the gas phase reaction mechanism, the carbonation reduction reaction of alumina is divided into several processes[3].[1]Migration of material through a gas layer on the surface of a particle.[2]Material migration of stomata between particles within particles.[3]Reduction and nitride of alumina particles mixed with CO-N_2gas; CO/N_2 mixture diffuses through the AlN layer on the particle surface; Diffusion of CO_2 gas products and reactions at the interface.[4]CO_2 is reduced by carbon.CO and N_2 diffuse inward through the product layer. The release of CO_2 products controls the reaction process.

The chemical reaction formula of the gas reaction mechanism is:

$$Al_2O_3 \rightarrow 2Al(g) + O_2(g)$$

$$Al_2O(g) + N_2 \rightarrow 2AlN$$

For the solid phase reaction mechanism, another reaction modelwas proposed by H K Chen[4] et al. The reaction is as follows.[1]Al_2O_3 and C first undergo a solid-solid reaction, producing Al_2O and CO gases.[2]Al_2O is adsorbed on the surface of Al_2O_3 and reacts with CO and N_2 to form AlN nucleus and generate CO_2[3]Once the AlN nucleus is formed, it begins to grow. At this time, the gas can be adsorbed on the surface of Al_2O_3.ThenAlN nucleus formed. [4]CO_2 which was formed in the reactionis then absorbed to the carbon and reacts with it to produce CO.

The chemical reaction formula of the solid phase reaction mechanism is:

$$Al_2O_3 + 2C \rightarrow Al_2O(g) + 2CO$$

$$Al_2O(g) + C + N_2 \rightarrow 2AlN + CO$$

According to different experimental conditions, There are problems with the mechanism of carbon-thermal reduction whether is explained by gas-solid reaction mechanism or solid-solid reaction mechanism.It is difficult to explain a specific carbon-thermal reduction reaction perfectly by a single mechanism, so it is proposed that two reaction mechanisms coexist.

Wengluqian[5] et al. believed that the zero-order reaction mechanism controlled by the evaporation of alumina exists simultaneously with the diffusion control of the solid reaction mechanism. Therefore, we believe that there are two reaction mechanisms in different stages of the carbon-thermal reduction reaction, which is more reasonable than other single reaction mechanisms.

3. Experimental materials and methods

3.1 Experimental materials

The details of materials used in the experiment are shown in Table 1.

Table 1. Experimental materials

Reagent	Molecular formula	Specification	Manufacturer
Ammonia	$NH_3 \cdot H_2O$	A.R	Shenyang chemical reagent factory
Ammonium Bicarbonate	NH_4HCO_3	A.R	Tianjin kemeiou chemical reagent development center
Polyvinyl alcohol	PEG4000	A.R	Shenyang chemical reagent factory
Aluminum nitrate	$Al(NO_3)_3 \cdot 9H_2O$	A.R	Shenyang chemical reagent factory
Ethanol	C_2H_5OH	A.R	Shenyang economic and technological development zone reagent factory
Deionized water	H_2O		Laboratory homemade
Sucrose	$C_{12}H_{22}O_{11}$	A.R	Tianjin ruijin chemical co. LTD

3.2 The experimental process

Synthesis of AlN precursor by mechanical hybrid method: Alumina and activated carbon prepared by chemical precipitation method are used as raw materials. The molar ratio of carbon to aluminum is C/Al=6. Ethanol was used as the ball grinding medium for wet grinding. The ball grinding speed was set as 1000r/min, the ball material ratio as 5:1, and the ball grinding time was 36h.After the mixture is ball-ground, the AlN precursor can be obtained. The aluminum nitride powder is obtained after the carbon removal of the precursor.

A new chemical reaction mechanism for the synthesis of aluminum nitride by carbothermal reduction of aluminum oxide is as follows:

$$Al_2O_3+N_2 \rightarrow AlON$$

$$AlON+C+N_2 \rightarrow AlN+CO$$

The route diagram of the synthesis process is shown as Figure 1.

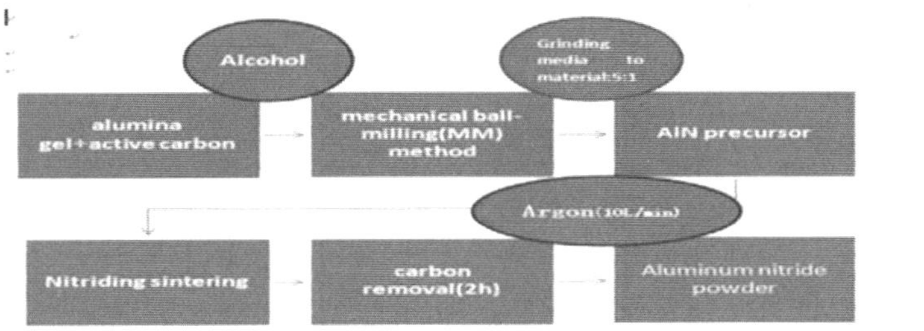

Figure 1. The flow sheet of AlN powder preparation

4. Synthesis of aluminum nitride powders

The chemical reaction equation of AlN powder prepared by carbothermal reductive nitriding is shown in equation (1).

$$Al_2O_{3(s)} + 3C_{(s)} + N_{2(g)} \rightarrow 2AlN_{(s)} + 3CO_{(g)} \tag{1}$$

Calculate the free change energy of the reaction, and then determine the temperature at which the reaction starts spontaneously. It can be found from literature [6] that: when the reaction that can occur at about 1400 °C. According to the theoretical calculation, the research groupaluminum nitride powder was calcined under the four groups of samples such as 1450°C, 1500°C, 1550°C and 1600°C respectively, and studied the reaction products at different temperatures.

4.1 Effect of synthesis temperature on the content of AlN powders

Table 2. Effect of synthesis temperature on synthesis powder by mechanical mixing method

Number	Temperature/°C	Heating time/h	O wt%	C %
1	1450	2	1.28	0.62
2	1500	2	1.11	0.59
3	1550	2	0.93	0.55
4	1600	2	0.75	0.48
5	1450	1	2.20	0.79
6	1500	1	1.70	0.78
7	1550	1	1.20	0.74
8	1600	1	1.36	0.71

Effect of synthesis temperature on synthesis powder by mechanical mixing method is shown as Table 2. When the insulation time was increasing, the oxygen content and carbon content of the products decreased under unchanged synthesis temperature. It indicated that the 1h insulation reaction time

IOP Conf. Series: Materials Science and Engineering **647** (2019) 012007 doi:10.1088/1757-899X/647/1/012007

cannot make the products completely nitrify. And the carbon content was getting smaller as the temperature rises up.

4.2. Effect of calcining temperature on phase composition of ALN powders

The XRD diagram of the reaction products of aluminum nitride precursor synthesized by mechanical mixing at different temperatures is shown as Figure 2. Aluminum nitride precursor was synthesized by mechanical mixing method, the AlN phasewas formed under 1450 °C, some Al_2O_3 are also exist in the products as can be seen from Figure 2 There are still a small amount of Al_2O_3 in the reaction product under 1500 °C and 1550 °C.But at 1600 °C AlN phase and a small amount of carbon were exist in the reaction product without any Al_2O_3 phase.

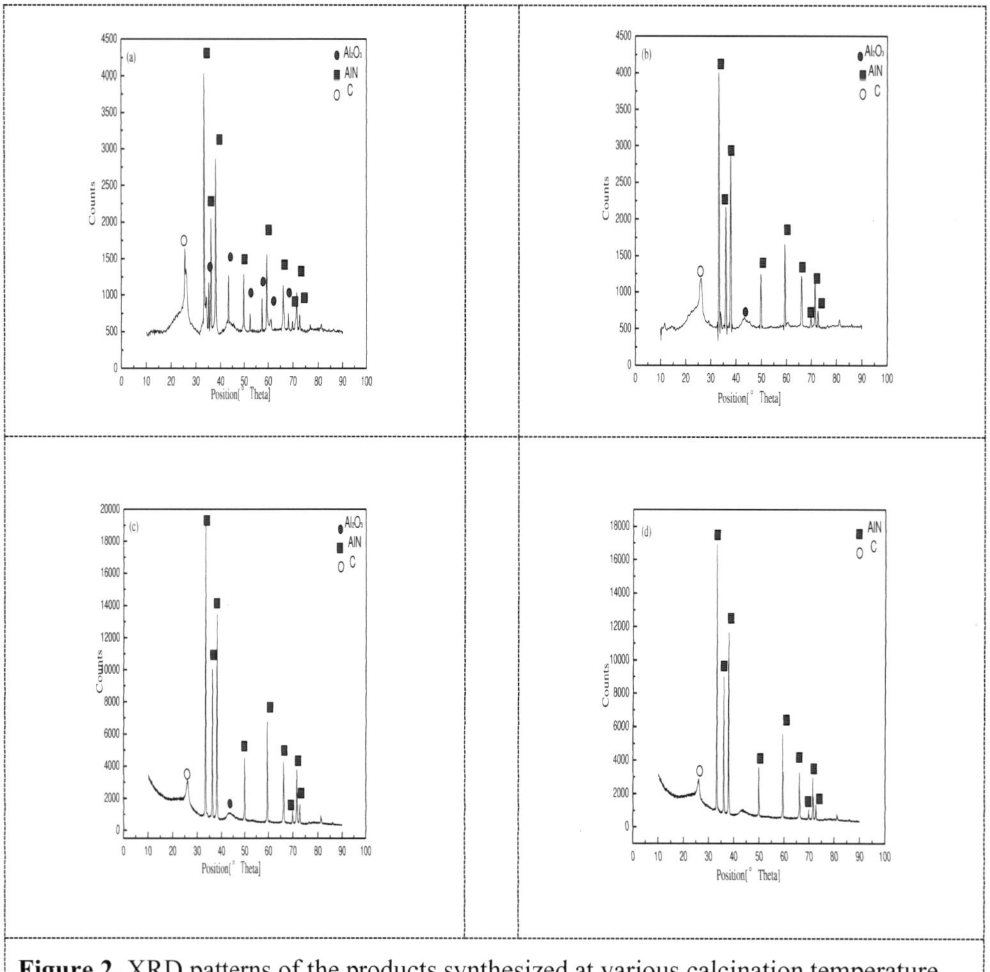

Figure 2. XRD patterns of the products synthesized at various calcination temperature (a)1450°C (b)1500°C(c)1550°C(d) 1600°C

Aluminum nitride precursor which prepared by ball milling method is calcined at 1600 °C for 2h under flowing nitrogen (10L/h).Aluminum nitride powder was finally synthesized after the powder was carbonized by 650°C for 4h.The final aluminum nitride powder has only a single aluminum nitride phase and its purity is high as shown in Figure 3.

As can be seen from Figure 2, the synthesized powder had secondary AlON phase besides main AlN phase before 1600°C, It indicates that there are a large number of AlN synthesis at this temperature, but the reaction is not complete. At the same time, it can be seen that with the increase of temperature, the characteristic peak of AlON crystal phase decreases until it disappeared.

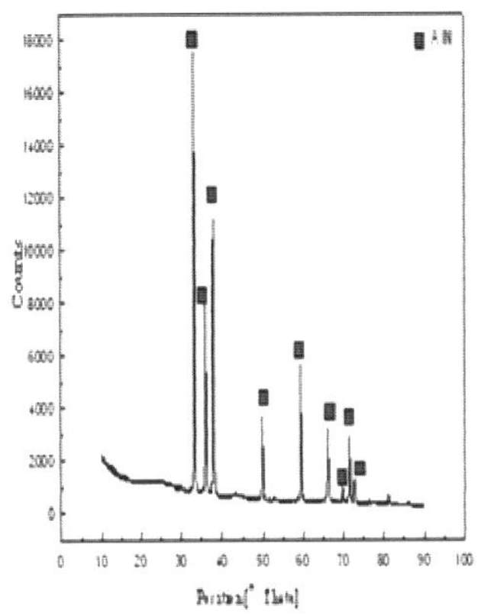

Figure 3. XRD patterns of aluminum nitride prepared by mechanical synthesis of aluminum nitride precursor.

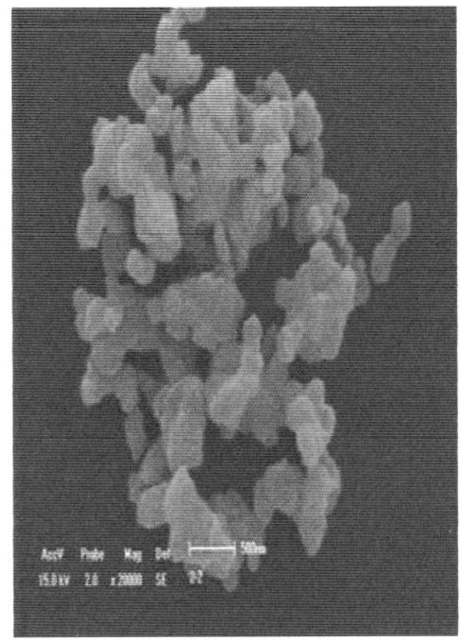

Figure 4. The scanning electron microscope image of aluminum nitridide powder.

4.3 Effect of calcining temperature on microstructure of ALN powders

Taking the aluminum nitride precursor synthesized by mechanical mixing method as raw material, calcining and holding for 2h under flowing nitrogen at 1600°C at 10L/h, the prepared powder was subjected to 650°C and kept carbonized for 4h before the final synthesis of aluminum nitride powder.

Figure 4 shows the scanning electron microscope image of aluminum nitridide powder synthesized by holding it at 1600°C for 2h with a magnification of 20,000 times. It can be seen from Figure 4 that the grain grains are relatively complete and clear in profile after 2 hours of insulation, indicating that the crystallinity is good under this process condition, which is conducive to the growth of grains.

5. Conclusion

According to the mechanism analysis of the carbon-thermal reduction method combined with the theoretical numerical experiment analysis, it is concluded in the experiment that aluminum nitride powder with good crystallinity and high purity can be obtained by keeping the temperature at 1600°C for hours.

Acknowledgments

The project was supported from Natural science foundation of Liaoning province (2018011155-301) and key laboratory of environment and geotechnical of Liaoning province. The authors thank the support of Northeastern University and Centre for Microscopy, Characterization and Analysis.

References

[1] R L Fu, H Z Hong, Y M Zhang, etal. Mechanism and thermodynamic conditions of synthesis of aluminum nitride by carbon thermal reduction [J]. journal of china university of mining & technology, 2002, 31(5):415-420.

[2] Y M Lu, T qiu, C Y shen, et al. Synthesis of aluminum nitride by carbon thermal reduction [J]. Jiangsu Ceramics, 2002, 35(3):4-6.

[3] X K Liu, M L Ma, S Q Xi, et al. Mechanism and thermodynamics of carbon thermal reduction of alumina [J]. T NONFERR METAL SOC, 2000(S1):46-49.

[4] Chen HK, Lin C I, Lee C J, J.Am.Ceram.Soc.,1994,77:1753-1756.

[5] F X Weng, J Chin Ceram Soe, 1993, 5: 29-31.

[6] Pankratz L B, et al. Thermodynamics Data for Mineral Technology United States Department of the Inferier Bueau of Mines, 669(1984).

IOP Conf. Series: Materials Science and Engineering **647** (2019) 012008 doi:10.1088/1757-899X/647/1/012008

Scaling Behaviour, Optical and Photoluminescence Properties of La_2Hf_2O Films with Annealing Treatments

Z Li[1, a], C Y Ma[2, b] and Q Y Zhang[2]

[1] Department of Mechanical Engineering, Dalian University, Dalian 116622, People's Republic of China

[2] Key Laboratory of Materials Modification by Laser, Ion and Electron beams (Dalian University of Technology) Ministry of Education, Dalian 116024, People's Republic of China

Corresponding author Email: [a] lizhi@dlu.edu.cn; [b]chunyuma@dlut.edu.cn;

Abstract. The structure, optical and PL characteristics of $La_2Hf_2O_7$ (LHO) films deposited by rf sputtering and subsequently annealed at 800-950°C are studied by x-ray diffraction (XRD), atomic force microscopy (AFM), UV/VIS spectrophotometry, and photoluminescence (PL) measurements. Structural studies indicate that LHO films with 37% atomic ratio of La/(La+Hf) have shown high crystallization temperature higher than 900°C. The scaling behaviors of AFM topographical profiles were analysed using one-dimensional power spectral density analysis (1DPSD). For as-grown films, only one growth exponent α is evidenced. For annealed films two different slopes α_1 and α_2 were observed, indicating distinct growth dynamic in LHO films. The estimated optical band gap was found to be in the range of 5.18-5.55 eV. Photoluminescence (PL) spectra of LHO films in visible light range (i.e., at 412~435 nm and 468 nm, respectively) can be attributed to localized states at the band tails and oxygen vacancies, respectively. This PL can be diminished or decreased by adjusting structural properties i.e. the amount of oxygen vacancies or some other kind of structure defects in the films by thermal annealing processing.

1. Introduction

$A_2B_2O_7$ complex oxides（A is a rare earth cation, B is a transition element such as Hf,Zr,Ti）attract plenty of interests due to their excellent thermo-mechanical stability, good catalytic efficiency, high ionic conductivity, unique magnetic properties, high efficiency as luminescence hosts, along with a strong resistance to amorphization under irradiation. These suitable properties render $A_2B_2O_7$ materials potentially good prospect for applications such as thermal barrier coating[1,2], electro- and photo-catalysts[3], magnetic materials[4], solid electrolytes for high temperature oxide fuel cells[5], defect-induced fluorescent materials and neutron absorbers in nuclear reactors[6,7], etc.

There exist in two major structures of $A_2B_2O_7$ complex oxides: the ordered pyrochlore having Fd-3m space group and the disordered fluorite with Fm-3m space group, which depends on ionic radius ratio of r_A and r_B(r_A/ r_B). According to Subramanian et al.[8], pyochlore structure steadily exists when r_A/ r_B =1.46~1.78. The $A_2B_2O_7$ pyrochlore structure mainly belongs to long range order of cationic positions and anionic vacancies, where the A sites are occupied by trivalent cations located in the centre of distorted cubes of oxygen anions, and the B sites are occupied by tetravalent transition metal ions situated in the centre of oxygen octahedral [9].

Interestingly, $La_2Hf_2O_7$ ($r_{La}^{3+}/r_{Hf}^{4+}=1.63$) usually existing in just one crystal form of the pyrochlore structure has been considered for the application of novel density radiation detectors due to a high density (7.84g/cm^3) and large effective atomic number (64)[10-12]. Meanwhile, amorphous $La_2Hf_2O_7$ is widely studied as a promising high-k candidate because of its rather lower defect densities, higher dielectric constant (20~30) and better thermal stability [13, 14]. Here, we focus on the amorphous and pyrochlore forms of $La_2Hf_2O_7$ films. Several methods have been explored for the formation of $La_2Hf_2O_7$ films, such as magnetron sputtering [14, 15], atomic layer deposition [16], metal organic chemical vapour deposition [17], pulsed laser deposition and molecular beam epitaxy [13, 18], etc. Among these methods, the sputtering method is a very useful technique because it provides large area coatings with good thickness uniformity, high film quality and strong mechanical durability.

Despite the wide interest for amorphous and pyrochlore $La_2Hf_2O_7$, the detailed study on structure, optical and PL characteristics of $La_2Hf_2O_7$ films due to the annealing process has not yet been reported, In this work, we report the observation of room temperature photoluminescence (PL) of $La_2Hf_2O_7$ films in visible light range (i.e., at 412~435nm and 468nm, respectively) due to localized states at the band tails and oxygen vacancies during deposition and possible approaches to decrease or diminish the PL of the films by adjusting structural properties i.e. the amount of oxygen vacancies or some other kind of structure defects in the films by thermal annealing processing. Also, the dynamics of the evolution of the surface morphology due to the annealing process was analysed using the one-dimensional power spectral density method (1DPSD).

2. Experimental details

$La_2Hf_2O_7$ (LHO) films were deposited onto n-type Si (001) wafers and optical grade quartz substrates by rf magnetron sputtering in an Ar/O_2 atmosphere. All the substrates were carefully cleaned by ultrasound in sequence in acetone, ethanol, and de-ionized water baths. N-type Si wafers were HF dipped to remove the native silicon oxide layer immediately before deposition, followed by rinsing in de-ionized water and dried under flowing N_2 gas. The vacuum chamber was evacuated to 8.0×10^{-4} Pa using a turbomolecular pump. LHO films were deposited by the co-sputtering of La and Hf targets (60 mm diameter, 3 mm thick and purity higher than 99.99%). The two disc-shaped targets are inclined toward the substrate at an angle of about 45°. The target to substrate distance is about 80 mm, and the substrate is rotated at 3 rounds / min for uniform film deposition. The flow of the Ar and O_2 and their ratio (30:5) was controlled using standard mass flow meters. Before each deposition, the Hf and La targets were presputtered for 10 min using Ar alone with shutter above the targets closed. The RF power of La and Hf targets was fixed at 100 W and 120 W, respectively. The deposition was carried out for 2 hours with sputtering pressure of 0.5 Pa. The substrate was not heated intentionally. Pure HfO_2 films were also prepared for comparison. Post-deposition annealing (PDA) was performed in air from 850 to 950°C for 30 min.

The composition of LHO films is measured by an energy-dispersive x-ray (EDX) spectrometer. The crystalline structure of the films was analysed by x-ray diffraction (XRD) using unresolved Cu Kα radiation (wavelength $\lambda_{x\text{-}ray} = 0.1542$ nm, D/Max 2400). A typical 2θ scan range of 25-80° and a step size of 0.02° were employed. Atomic force microscopy (AFM) on a CSPM5500 scanning probe microscope was used in contact mode with a scan frequency of 1.0 Hz to measure surface morphologies and roughness of the films. The measurement of transmittance (T) and reflectance (R) at normal incidence of the films deposited on quartz substrates were performed in the wavelength range of 190-1100 nm using a Lambda-35 UV/VIS spectrometer. Photoluminescence (PL) spectra were recorded by using a Fluorescence spectrometer (Edinburgh Instr.Model-FLS920) with a 450W Xenon lamp as excitation source.

3. Results and discussion

3.1. X-ray diffraction

Figure 1 shows XRD patterns for the LHO films with as-grown and after PDA in air from 850 to 950°C for 30 min. EDX analysis (not shown) of the LHO films gave an estimation of 0.37:1 for the atomic ratio of La /(La+Hf). A typical XRD pattern of as grown HfO_2 films is also shown. Generally,

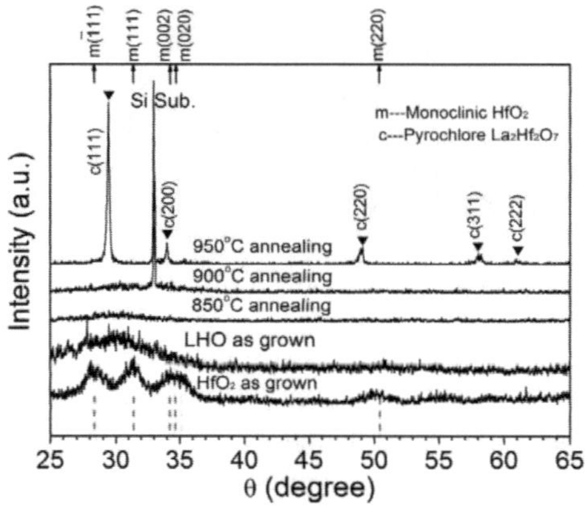

Figure 1 XRD patterns of LHO films.

HfO_2 and La_2O_3 films start to crystalline below 600°C [19, 20], while the incorporation of La in periodic HfO_2 shows higher crystallization temperature than pure HfO_2 or La_2O_3. In Figure 1, the XRD curve of the pure HfO_2 film indicates the polycrystalline nature of samples. The peaks can be unambiguously assigned to monoclinic HfO_2 as labelled in Figure 1 (referenced JCPDS 74-1506). It is clear that as-grown LHO films are amorphous and remain amorphous up to 900°C. This is probably due to the La acting as a network modifier, which stabilizes the amorphous nature of its metal oxide. It is also believed that the incorporation of La breaks the periodicity and/or inhibits the continuous crystal growth of HfO_2 during PDA. Increasing the annealing temperature to 950°C, LHO films crystalline to be the cubiclike pyrochlore structure, and this is clearly distinguished from the monoclinic structure of HfO_2. A diffraction peak around $2\theta = 29.4°$ has been observed, which is the strongest diffraction peak of pyrochlore structure having Fd-3m space group, corresponding to the (111) planes. The other four weak diffraction peaks at 34.1°, 49.1°, 58.4° and 61.2° can also be observed, corresponding to the (200), (220), (311) and (222) planes of $La_2Hf_2O_7$, respectively. This indicates a polycrystalline LHO film with predominant (111) orientation on Si (100) is formed after annealing at 950°C, suggesting that surface and interfacial energy minimization may provide an additional driving force for the preferential growth of the (111) oriented grain.

3.2. Surface morphology and scaling analysis

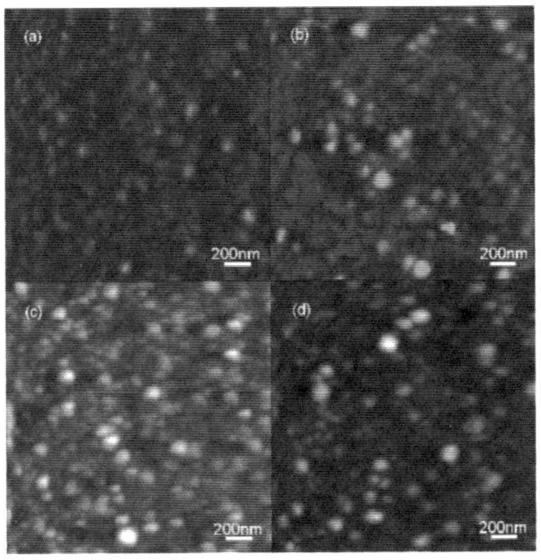

Figure 2 AFM images with Z=20 nm of LHO films: (a) as-grown, (b) annealed at 850°C, (c) annealed at 900°C and (d) annealed at 950°C.

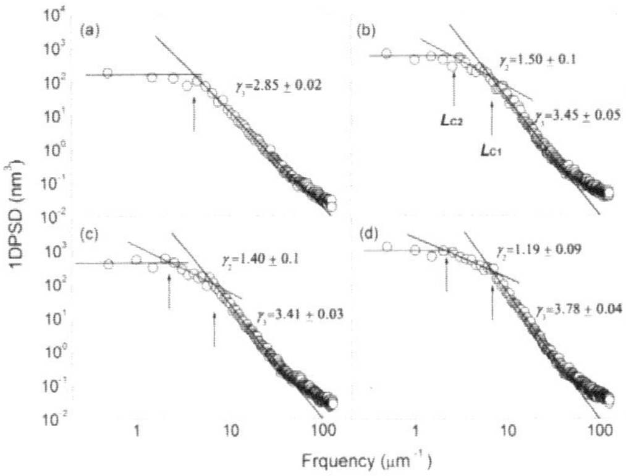

Figure 3 Double logarithmic plots of 1DPSD for LHO films: (a) as-grown, (b) annealed at 850°C, (c) annealed at 900°C and (d) annealed at 950°C.

Figure 2 shows AFM images of the LHO films prepared by rf magnetron sputtering on Si (100) wafers. By visual inspection of these figures two morphological features are recognized. The first feature is that the granules of various scales exist in the films and are distributed evenly in some ranges. The second feature is that the evolution of root mean square roughness (rms) as a function of annealing temperature up to 950 °C, where valleys, mountains, and island clusters become bigger as films are annealed. The results (*R*a and rms) have been listed in Table 1. It can be see that the average roughness *R*a and rms values increased as the annealing temperature is increased. The above analyses indicate average roughness *R*a and rms are strongly affected by the degree of aggregation and cluster size of the films.

The surface morphology has an irregular geometry with self-affine structure determined by the thermodynamic environment. Power spectral density (PSD) analysis has been widely used for quantitative surface characterizations with spatial resolution of the roughness and correlations between roughness and specific surface features [21-23]. We analyse the scaling behaviour of the AFM topological profile using the one-dimensional power spectral density (1DPSD).

$$1DPSD(f) = \frac{1}{L}\left|\int_0^L y(x)e^{i2\pi fx}dx\right|^2 \quad (1)$$

in unites of nm^3, where $y(x)$ is the topographical profile, L is the scan length, f is the spatial frequency, and x is the fast scan direction of an area scan. The scaling behaviour of surfaces can be analysed from their power density spectra. In a finite frequency range, a self-affine structure exhibits the power law decay

$$1DPSD = K_0 f^{-\gamma} \quad (2)$$

where K_0 is a constant dependent on the system and γ is related to the roughness exponent (α) by $\alpha = (\gamma - d)/2$, where the line scan dimension $d = 1$.

The log-log 1DPSD plots of AFM topographical morphologies corresponding to the as grown and annealed samples in Figure 2 are presented in Figure 3. The 1DPSD plots exhibit several distinct regions, a frequency-independent region at low frequency (long length scales), and a frequency-dependent region at higher frequency (short length scales). Interestingly, for the annealed LHO films, Figure 3 (b)-(d), this higher frequency behaviour is composed of two constant-slope regions with different roughness exponents. The dependence can be fitted by $f^{-\gamma_3}$ and $f^{-\gamma_2}$ power law decays at high and intermediate frequencies, respectively. According to the scaling theory, the appearance of an intermediate region (- γ_2) indicates a new coarsening mechanism, such as grain growth and oriented growth effects.

The roughness exponents were extracted from the two constant- slope 1DPSD regions based on Eq. (2). As listed in Table 1, the only roughness exponent α_1 for as grown LHO films is about 0.93, while two roughness exponents with α_1 (the higher-frequency region) and α_2 (the intermediate-frequency region) for our annealed LHO films are found. In addition, for amorphous LHO films annealed at 850-900°C, the two roughness exponents α_1 and α_2 are nearly constant, with values in the range 1.21~1.23 and 0.2~0.25, respectively. With the films annealed at 950 °C presenting a clear transition from amorphous to crystalline structures, α_2 decreases from 0.2 to 0.1, while α_1 increase from 1.21 to 1.39.

These roughness exponents are similar to previous theoretical results that have identified two types of growth front dynamic scaling behaviour. The first type has a self-affine growth morphology characterized by a roughness exponent $0<\alpha<1$. This behaviour has been described as a balance between roughening mechanisms, such as random fluctuations in the depositing flux, and smooth processes, such as diffusion at the growth front so that the local structure remains unchanged. The second type is anomalous dynamic scaling, which gives $\alpha\geq1$. In this situation the roughening fluctuations and the smooth effects cannot quite reach a balance. We suggest that for our annealed LHO films, the intermediate-frequency self-affine region with α_2 = 0.1~0.25 is characterized by balance between grain and oriented growth induced roughening and longer range diffusion processes so that the local structure remains unchanged, while the high frequency region with α_1 =1.21~1.39 corresponds to anomalous dynamic scaling behaviour dominated by surface diffusion.

The frequencies of transition marked by arrows in Figure 3 between the various regimes of 1DPSD behaviour are of interest since they infer some physically meaning correlation lengths. The first correlation length (L_{C1}) is defined by the (inverse) frequency of transition between the high-frequency and intermediate-frequency regions, which marks the beginning of the surface diffusion part of $1DPSD=K_0f^{-\gamma}$. For annealed LHO films, the L_{C1} values (about 150 nm) extracted at points of 1DPSD slope change are approximately constant which represents the minimum lateral feature stable against surface diffusion. Surface features with lateral dimensions smaller than those defined by L_{C1} will gradually vanish by surface diffusion. Similarly, the second correlation length (L_{C2}) indicates the transition between the low-frequency plateau and the intermediate-frequency self-affine region in the 1DPSD, which therefore defines the mean lateral extension of a surface feature stable against surface diffusion. L_{C2} values for LHO films are also listed in Table 1, it can be seen that L_{C2} values increase with increasing annealing temperature.

Table1. Surface roughness (R_a and rms) and morphological characteristics from analysis of 1DPSD plots for LHO films.

Sample		As-grown	Annealing temp.850℃	Annealing temp.900℃	Annealing temp.950℃
AFM	R_a(nm)	0.75	1.06	1.23	1.70
	rms(nm)	1.01	1.49	1.64	2.28
	L_{C1}(nm)	-	148	150	149
	L_{C2} (nm)	242	383	452	465
	γ_2	-	1.50±0.1	1.40±0.1	1.19±0.09
	γ_3	2.85±0.02	3.45±0.05	3.41±0.03	3.78±0.04
	α_1	0.93	1.23	1.21	1.39
	α_2	-	0.25	0.20	0.10

3.3. Optical characterizations

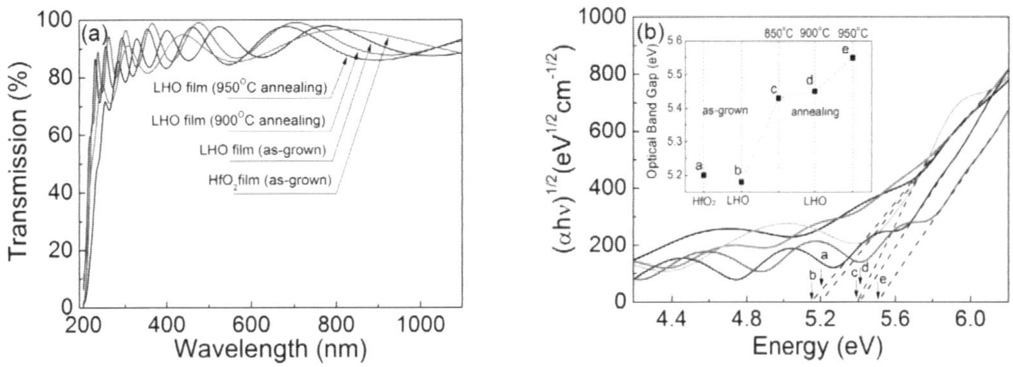

Figure 4 (a) Transmittance spectra of as grown and annealed LHO films on quartz substrates. (b) Plots of $(\alpha h\nu)^{1/2}$ versus $E(h\nu)$. Inset shows the variation in optical band gap.

Optical transmission spectra of the films were measured using an UV/visible spectrophotometer. As shown in Figure 4 (a), All LHO films exhibit very high transparency in the visible optical region. LHO films exhibits higher transmittance than HfO_2 films. This observation indicates the high quality and transparent nature of LHO films with almost zero absorption losses. Moreover, it can be seen that the transmittance maxima of amorphous LHO films annealed at 900 °C are slightly higher than that of pyrochlore LHO films annealed at 950 °C. In comparison with that of pure HfO_2 film, it can be found that the absorption edge shifts to the shorter wavelength for annealed LHO, which is correlated to the change in the optical band gap value. The optical absorption coefficient of the films is evaluated using the relation:

$$\alpha = \frac{1}{t}\ln\left[\frac{T}{(1-R)^2}\right] \tag{3}$$

where T is the transmittance, R the reflection, and t the film thickness. In the E_g region (high absorption) or above the fundamental absorption edge, the absorption follows a power law of the form

$$(\alpha h\nu) = B(h\nu - E_g)^2 \tag{4}$$

where $h\nu(E)$ is the energy of the incident photon, B the absorption edge width parameter, and E_g the band gap. The $(\alpha h\nu)^{1/2}$ versus E plots for LHO films are shown in Figure 4 (b). The linear fit of the data indicates the characteristic feature of LHO with an indirect band gap. Extrapolation of the linear region of the plot to zero absorption i.e., $(\alpha h\nu)^{1/2} = 0$, provides the value of E_g. The extracted band gap of as grown LHO is 5.18 eV, which is close to the band gap value 5.2 eV of pure HfO_2 film. The band gap energies of 850 °C and 900°C annealed LHO films are 5.43eV and 5.45eV, respectively, and that of pyrochlore LHO films annealed at 950°C increases to 5.55 eV.

An increase in the optical band gap after annealing at 850-900 °C compared to that of as grown LHO is obtained, which can be attributed to a decrease in the density of tail adjacent to the band edge. It is known that dangling bands are formed in amorphous films during deposition and they will lead to the formation of defects which produces localized states in amorphous solids. During the heat treatment, a great fraction of saturated bands are produced. The formed saturated bonds will reduce the number of defects and decrease the density of localized states in the band structure. On the other hand, this study confirms that band gap energy of the pyrochlore LHO films with more long range ordering is larger than that of the amorphous LHO film. XRD measurement clearly indicates that grain growth on LHO films annealed at 950°C. The observed blue shifting of band gap in pyrochlore LHO films can be mainly ascribed to the contribution of quantum confinement.

3.4. Photoluminescence characterization

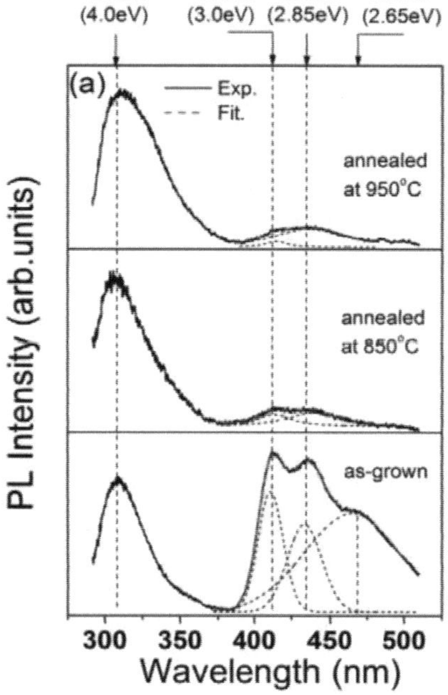

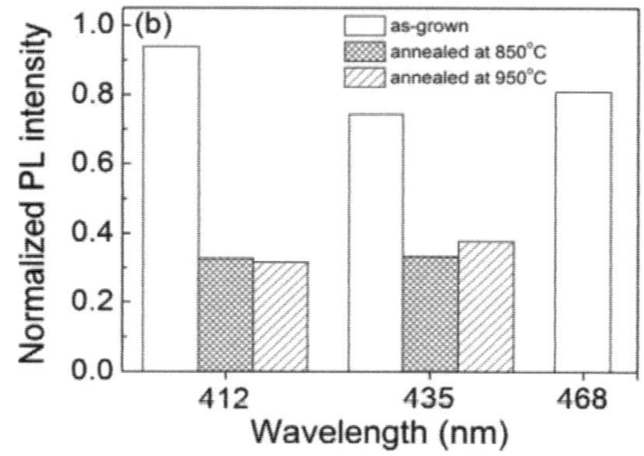

Figure 5 (a) Photoluminescence spectra of for LHO films. (b) Changes in PL intensity around 412 nm, 435 nm and 468 nm. Note that the PL intensity was normalized by the intensity around 308 nm.

Photoluminescence (PL) is a very sensitive and effective tool for the investigation of electron structure of defect centers [24, 25]. Figure 5(a) shows the PL spectra recorded at room temperature for LHO films as grown and after annealed at 850 °C and 950°C, respectively, using photoluminescence excitation of 280 nm. A broad ultraviolet PL signal is present with a peak value at 308 nm (4.0 eV) in all LHO samples. This PL is considered to be excited due to the interb and absorption. In addition to the ultraviolet PL peak, visible PL peaks in the range 468 nm (2.65 eV) to 412 nm (3.0 eV) are observed for the case of the as grown sample, which can be fitted based on three Gaussian peaks PL peaks at 468 nm(2.65 eV), 435 nm(2.85) and 412 nm(3.0 eV).

The measured intensity of a PL peak generally depends on various hardly controllable factors, such as a subtle difference in the mutual angles among the photon source, the sample, and the detector. This sometimes makes it difficult to compare the PL intensities among different samples. Therefore, the PL intensities around 3.0 eV, 2.85 eV and 2.65 eV are shown in Figure5 (b) after being normalized by the highest peak around 4.0 eV. It is noted that the PL peak at 468nm (2.65eV) is not present in the thermally annealed samples. $La_2Hf_2O_7$ crystallizes in the cubic pyrochlore structure. The structure can be viewed as an interpenetrating network of HfO_6 and LaO_8 polyhedra whose geometries are dependent upon the amount of oxygen in the lattice [26]. A broad emission band peaking at about 468 nm found in a spectrum of as grown LHO films can be attributed to oxygen vacancies. The additional air annealing may enable oxygen to fill in the existing vacancies. Decreasing the concentration of O vacancies with in LHO films will thus lead to a decrease or disappear in the PL intensity. These results are indicative of the formation of more oxygen vacancies during the film deposition.

The PL peaks at 412nm (3.0eV) and 435nm (2.85) are present in common with all as grown (amorphous) and annealed (amorphous or crystalline) samples. The as grown and annealed samples have crystallinity from amorphous to crystalline. However, the difference in crystallinity is not reflected in the PL properties. As mentioned above, oxygen vacancies are certainly not responsible for the present PLs. Furthermore, the thermal treatment decreases their PL intensities as shown in Figure 5 (b). It seems that some structural change induced by the thermal treatment extinguishes localized states at the band tails due to defects, although the details are unknown. Therefore, The PL peaks at

412 nm (3.0eV) and 435 nm (2.85) are luminescence inherent in LHO films, and are most likely due to radioactive recombination between localized states at the band tails.

4. Conclusion

In summary, we investigated the structure, scaling behaviour, optical and PL characteristics of LHO films. LHO films have shown high crystallization temperature (＞900°C). Quantitative surface characterization for the films in the annealing process by 1DPSD analysis identified two types of growth front dynamic scaling behavior. For annealed LHO films, the intermediate-frequency self-affine region with $\alpha_2 = 0.1\sim0.25$ is characterized by balance between grain growth induced roughening and longer range diffusion processes, while the high frequency region with $\alpha_1=1.21\sim1.39$ corresponds to anomalous dynamic scaling behavior dominated by surface diffusion. Photoluminescence (PL) spectra show that for as grown LHO films a broad emission band peaking at about 468 nm can be attributed to oxygen vacancies, indicating the formation of more oxygen vacancies during the film deposition. In addition, the PL peaks at 412 nm (3.0 eV) and 435 nm (2.85eV) are luminescence inherent in LHO films, and are most likely due to radioactive recombination between localized states at the band tails. This PL can be diminished or decreased by adjusting structural properties i.e. the amount of oxygen vacancies or some other kind of structure defects in the films by thermal annealing processing.

Acknowledgments
The work is supported by the Fundamental Research Funds for the Central Universities no. DUT19LAB14.

References

[1] Bansal N P and Zhu D M 2007 Mater. Sci. Eng. A45 192-195
[2] Zhang J, Guo X Y, Jung Y G, Knapp, J 2017 Surf.Coat.Technol.323 18-29
[3] Yin X, Li X, Gu W, Wang F, Zou Y, Sun S, Fu Z, Lu Y 2017 ACS Appl. Mater. Interfaces 9 19908-16
[4] Fennell T, Deen P P, Wildes A R, Schmalzl K, Prabhakaran D, Boothroyd A T, Aldus R J, McMorrow D F, Bramwell S T 2009 Scinece 326 415-417
[5] Mahato N, Banerjee A, Gupta A, Omar S, Balani K 2015 Prog.Mater.Sci. 72 141-337
[6] Gupta S K, Sudarshan K, Ghosh P, Srivastava A, Bevara S, Pujari P, Kadam R 2016 J.Mater.Chem.C4 4988-00
[7] Ewing R C, Weber W J, Lian J 2004 J.Appl.Phys. 95 5949-71
[8] Subramanian M, Aravamudan G, Rao C S 1983 Prog.Solid State Chem.15 55-143
[9] Burggraaf A J, van Dijk T, Verkerk M J 1981 Solid State Ionics 5 519-22
[10] Zuniga J P, Gupta S K, Pokhrel M, Mao Y B 2018 New J. Chem. 42 9381-92
[11] Karthik C, Anderson T J, Gout D, Ubic R 2012 J.Solid State Chem. 194 168-172
[12] Nakauchi D, Okada G, Kawaguchi N, Yanagida T 2018 Jpn.J.Appl.Phys .57 100307
[13] Dimoulas A, Vellianitis G, Mavrou G, Apostolopoulos G, Travlos A, Wiemer C, Fanciulli M, Rittersma Z M 2004 Appl.Phys.Lett.85 3205-07
[14] Yamamoto Y, Kita K, Kyuno K, Toriumi A 2006 Appl.Phys.Lett.89 032903
[15] Wang X P, Li M F, Chin A, Zhu C X, Shao J, Lu W, Shen X C, Yu X F, Chi R, Shen C et al 2006 Solid-State Electron 50 986-991
[16] He W, Chan D S H, Kim S J, Kim Y S, Kim S T, Cho B J, 2008 J.Electrochem.Soc. 155 G189-G193
[17] Kaichev V V, Smirnova T P, Yakovkina L V, Ivanova E V, Zamoryanskay M V, Saraev A A, Pustovarov V A, Perevalov T V, Gritsenko V A 2016 Mater.Chem.Phys. 175 200-205
[18] Xiong Y H, Tu H L, Du J, Wei F, Zhang X Q, Yang M M, Zhao H B, Chen D P, Wang W W 2013 Appl.Surf.Sci. 283 554-558

[19] Triyoso D H, Hegde R I, Grant J, Fejes P, Liu R, Roan D, Ramon M, Werho D, Rai R, La L B et al 2004 J.Vac.Sci.Techol.B22 2121-27
[20] Ma J W, Lee W J, Cho M H, Lee K M, Sohn H C, Kim C S, Cho H J 2011 J. Appl. Phys. 109 124106
[21] Lita A E, Sanchez J E 1999 J. Appl. Phys. 85 876-882
[22] Li B Q, Kojima I, Zuo J M 2002 J. Appl. Phys. 91 4082-89
[23] Eisenmenger-Sittner C 2001 J. Appl. Phys.89 6085-91
[24] Papernov S, Brunsman M D, Oliver J B, Hoffman B N, Kozlov A A, Demos S G, Shvydky A, Cavalcante F H M, Yang L, Menoni C S 2018 Opt. Express 26 17608-23
[25] Rastorguev A A, Belyi V I, Smirnova T P, Yakovkina L V, Zamoryanskaya M V, Gritsenko V A, Wong H 2007 Phys. Rev.B 76 235315
[26] Subramanian M A, Aravamudan G, Rao G V S 1983 Prop. Solid State Chem.15 55-143.

IOP Conf. Series: Materials Science and Engineering **647** (2019) 012009 doi:10.1088/1757-899X/647/1/012009

Development of a Cutting Force Prediction Model for Silica/Phenolic Composite in Mill-grinding

Zhang Chong*, Zhang Tingyu, Wang Qi, Ling Li and Zhao Yibo

Aerospace research institute of materials & processing technology, Beijing, China

*E-mail: Zhangchongbuaa@163.com

Abstract. A cutting force model was developed from silica/phenolic composites for mill-grinding process. The experimental was carried out on silica/phenolic material and found that the cutting force decreased significantly with the increase of cutting speed, whereas the same was found increased with the increase of feed rate and cutting depth. By comparison of the experimental and simulation data of the cutting force, it was found that the errors are below than 30 % in most of the sets of parameters. The variation found is due to the heterogeneity and other complex properties of silica/phenolic composites. The results were almost the same as previous experiments. So, the cutting force model developed in this paper is robust and it can be applied to predict the cutting force and optimization of the process.

1. Introduction

Silica/phenolic composite are attractive due their superior properties such as low density, low thermal conductivity, ultra high-adsorption ability, ablation resistant performance and thermal stability. Presently, Silica/phenolic composite are used in thermal protection systems[1]. Their applications are increasing in space, military, and aerospace industries day by day. In traditional machining processes, the cutting tools wear very quickly and the cutting tools like polycrystalline diamond (PCD) tools are very expensive. It is challenging to achieve desired accuracy, high efficiency, and cost-effective processing for such composites. Keeping in view of their special properties and critical engineering applications, there is a crucial need to research for machining of silica/phenolic composites in order to achieve economic and efficient processing with desired quality. Low efficiency, high processing costs, and quality issues are three main problems which hindered the applications of silica/phenolic materials. With mill-grinding, better results have been achieved as compared to conventional machining of composites materials. Jinguang Du developed a cutting force mathematical prediction model on the basis of the study on material removal mechanism of SiCp/Al. Based on the cutting force model, the distribution of chip deformation component force, friction component force and particle fracture with the change of machining parameters, such as cutting speed, feed rate and milling width, was calculated[2]. Li Wanqing studied the effects of cutting speed, feed rate, and cutting depth on mill-grinding foece. The mill-grinding force has found increased with the increase of feed rate and cutting depth while it has decreased with the increase of cutting speed[3].

In this paper, the mechanistic model is developed to predict the cutting force for Silica/phenolic based on indentation fracture mechanics in mill-grinding. The mathematical relationship between cutting parameters (spindle speed, feed rate, and cutting depth) and cutting force is proposed. The parameter, k1, k2, k3, was obtained using response surface method. This paper is organized into five sections. After this introduction section, the predicted cutting force model is designed and developed in Sect. 2. In Sect. 3, the experimental mill-grinding is carried out and the data acquired is reported. The results and discussion are mentioned in Sect. 4. Finally, the conclusions are presented in Sect. 5.

2. Modeling

The material removal mechanism in mill-grinding is based on indentation fracture theory. When the diamond abrasive grit penetrates into the surface of the workpiece material, there will be a plastic deformation. With the increase of penetration depth, the median crack will grow and also generate the lateral crack. The extended lateral cracks then induce and peeling off the workpiece material occurs as shown in 'figrue 1'[4]. The maximum penetration depth (w) was used as an intermediate parameter to establish the relationships between the input parameters (spindle speed, feed rate, and cutting depth) and the output parameter, i.e., cutting force.

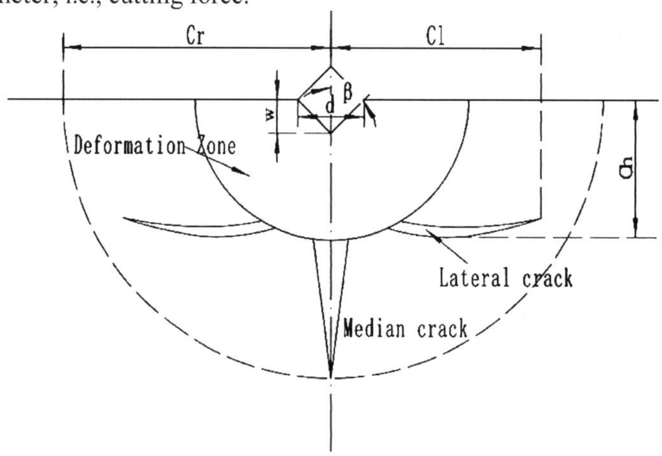

Figure 1. Crack propagation

Because of the randomness of distribution of wear debris, the abrasive angle is not a constant number. As shown in 'figure 2', while the value of penetration depth is w, the maximum and the minimum of the sweep area are $\triangle OBD$ and $\triangle OEF$, the area can be expressed as[5]:

$$S_{\max} = \frac{1}{2} \cdot 2w \cdot w = w^2 \tag{1}$$

$$S_{\min} = \frac{1}{2} \cdot \sqrt{2}w \cdot w = w^2 = \frac{\sqrt{2}}{2} \cdot w^2 \tag{2}$$

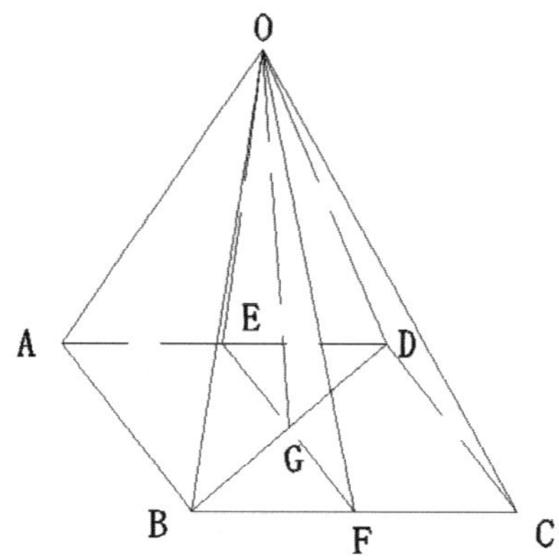

Figure 2. The sweep area of the diamond grits

The mean area was used to simplify the calculation. The average area can be calculated as follows:

$$S = \frac{1}{2}\left(S_{\max} + S_{\min}\right) = \frac{2+\sqrt{2}}{4} \cdot w^2 \tag{3}$$

The microscopic contact length of abrasive and materials is shown in 'figure 3'. The contact angle θ can be expressed as:

$$\theta = \arccos\left(1 - \frac{2w}{D}\right) \tag{4}$$

As $w<<D/2$, θ can be expressed as follows:

$$\theta = 2\sqrt{w / D} \tag{5}$$

The value of microscopic contact length can be expressed as:

$$l_c = \frac{D}{2} \cdot \theta = \sqrt{w \cdot D} \tag{6}$$

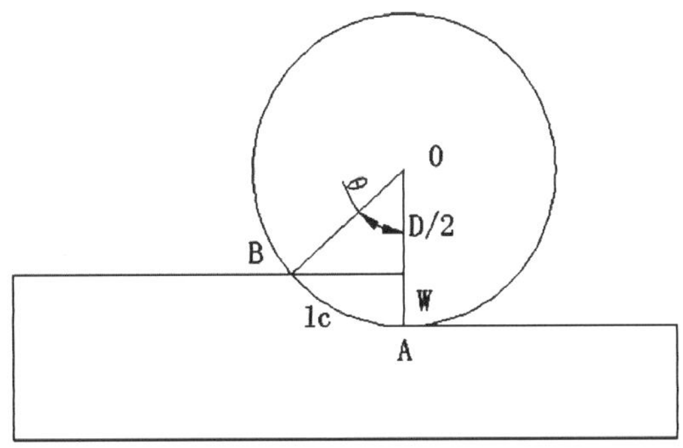

Figure 3. The microscopic contact length of abrasive and materials (l_c)

From a rather macroscopic perspective, the value of macroscopic contact length can be expressed as:

$$l_a = \frac{D}{2} \cdot \arccos\frac{D - 2a_e}{D} \tag{7}$$

Diamond abrasive concentration is the mass of abrasive per unit volume within working layer. Concentration is generally defined as follows: per cubic centimeter volume of abrasive grains containing 4.4 karats (1 karat diamond is equal to 0.2 g) is defined as 100. Each has an increase or decrease of 1.1 karats of abrasive, and then the concentration is increased or decreased by 25 %, respectively. According to this definition, the total number of active diamond abrasive grains involved in cutting, Nα, can be expressed as

$$N_\alpha = \left(\frac{0.88 \times 10^{-3} \times C_\alpha}{\frac{\sqrt{2}}{3} \times S_a^3 \times \rho} \cdot \frac{C_\alpha}{100}\right) \cdot A_0 = 0.03 \times \frac{C_\alpha^{2/3}}{S_a^2} \cdot l_a \cdot a_p \tag{8}$$

Material removal rate MRR is the summary of the material removed by all the effective abrasive particles during one period and can be expressed as follows:

$$MRR = N_\alpha \cdot V_c \cdot S \tag{9}$$

MRR can also be expressed as the volume swept by the cutting tool during one period:

$$MRR = a_e \cdot a_p \cdot f \tag{10}$$

By solving both Eqs. (9) and (10), the relationship between maximum penetration depth and cutting parameters can be obtained as follows:

$$w = \left(\frac{4 \times a_e \cdot f_r \cdot S_a^2}{0.03 \times \left(2 + \sqrt{2}\right) \cdot C_\alpha^{\frac{2}{3}} \cdot \frac{D}{2} \cdot \arccos \frac{D - 2a_e}{D} \cdot \sqrt{D} \cdot S} \right)^{\frac{2}{5}} \tag{11}$$

The cutting of mill-grinding can be expressed as three components: F_x, F_y. F_x is caused by cutting materials of abrasive particles directly, can be expressed as follows:

$$F_x = N_\alpha \cdot F_a = k_1 \cdot N_\alpha \cdot w^2 \cdot \sigma_b \tag{12}$$

Where F_a is the cutting force caused by one diamond abrasive, σ_b is the yield strength of the silica/phenolic composite, k_1 is a constant number which demarcated by experiment.

F_y is proportional to the cutting area per unit time, which can be express as follows:

$$F_y = k_2 \cdot 0.03 \times \frac{C_\alpha^{2/3}}{S_a^2} \cdot a_e \cdot a_p \cdot w^2 \cdot \sigma_b \tag{13}$$

Where k_2 is a constant number which demarcated by experiment.

3. Experimental setup

The schematic and the actual experimental setup is shown in 'figure 4', respectively. The setup is composed of three parts: three-axis milling machine (VMC0850B, Shenyang, China), diamond abrasive tool, and dynamometer (9257B, Kistler, Switzerland).

Silica/phenolic composite were used as the workpiece sample materials. The mechanical properties of Silica/phenolic composite are shown in Table 1. The parameters of diamond abrasive tool are shown in Table 2.

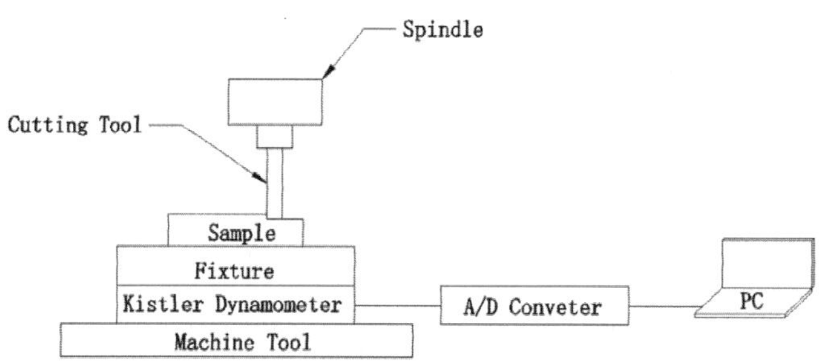

Figure 4. Experimental setup

Table 1. Mechanical properties of Silica/phenolic composite workpiece material

Density (g/cm^3)	1.6
Tensile strength(MPa)	20
Compression strength (MPa)	100
Bending strength(MPa)	50
Elongation at break(%)	0.35

Table 2. The properties of diamond abrasive tool

Abrasive	Diamond
Bond type	Electroplated

Grain size	60/70#
Concentration	$C\alpha$=100
Radius	R=16mm

The experimental design is shown in Table 3. Three groups of input parameters such as spindle speed, cutting depth, and feed rate were used in this research. These parameters were found significant on the basis of random experiments and the findings of previous research work. These cutting parameters were designed by response surface method with three factors.

Table 3. Experimental design

Factor	Level 1	Level 2	Level 3
Spindle Speed(rev/min)	1000	3500	6000
Feedrate (mm/min)	150	325	500
Cutting Depth(mm)	0.2	1.1	2

4. Experimental results and discussion

Table 4 shows the results of cutting force tests of cutting force. It was found that the simulation values are closest to measurement values, when $\Sigma\ (F - k{*}F_S')^2$ got the minimum value. Here, k_1 has obtained as 1.44, k_2 has obtained as 5.38. The comparative analysis of measured values and simulated values of cutting force was carried out, and finally, the percentage error was found for each set of experiments. It was found that in most of the cases, the error is below 30 % expt 16 (77 %). Such variations are mainly due to the heterogeneity of silica/phenolic.

Table 4. Comparative analysis of measured values and simulated values of Fx

	Spindle Speed rev/min	Feedrate mm/min	Cutting Depth mm	Fx N	Simulation F_x, N	Error F_x, %	Fy N	Simulation F_y, N	Error F_y, %
1	3500	500	2	164	173	5	250	265	6
2	3500	325	1.1	69	68	-2	97	103	6
3	3500	325	1.1	70	68	-3	100	103	3
4	1000	150	1.1	134	99	-26	215	151	-30
5	6000	325	2	86	79	-8	122	124	2
6	3500	325	1.1	63	68	7	92	103	12
7	6000	500	1.1	79	62	-22	111	92	-17
8	3500	325	1.1	70	68	-3	106	103	-3
9	1000	325	2	286	333	16	479	508	6
10	1000	325	0.2	37	33	-10	38	49	28
11	3500	325	1.1	68	68	0	98	103	5
12	1000	500	1.1	300	259	-14	412	394	-4
13	3500	150	2	78	66	-15	109	103	-6
14	6000	150	1.1	29	23	-21	50	36	-28
15	3500	150	0.2	8	7	-10	9	10	14
16	6000	325	0.2	11	9	-21	7	12	77
17	3500	500	0.2	17	17	2	22	26	20

5. Summary

In this research, mill-grinding was carried out on silica/phenolic composite. The following conclusions can be drawn:

1. The cutting force model was developed and then validated by experimental machining of silica/phenolic using mill-grinding. It was found that percentage difference/error between the measured and simulated (from model) values of cutting force is less than 30 % in the most set of values. However, in few cases, this difference was recorded more than 30 %. The cause of this difference is due to the heterogeneity and some other factors related to silica/phenolic composites. So, this model is robust and can be applied for finding cutting forces.

2. The cutting force has found increased with the increase of feed rate and cutting depth while it has decreased with the increase of spindle speed.

References

[1] Loomis M P, Prabhu D K, Gorbunov S, et al. Results and Analysis of Large Scale Article Testing in the Ames 60 MW Interaction Heating Arc Jet Facility[C]. 48th AIAA Aerospace Sciences Meeting Including the New Horizons Forum and Aerospace Exposition. Florida: Orlando, 2010-445.

[2] Jinguang Du, Jianguang Li, Yingxue Yao, Zhaopeng Hao. Prediction of cutting forces in mill-grinding SiCp/Al composites[J]. Materials and Manufacturing Processes, 2014, 29(3): 314-320.

[3] Li Wanqing. Optimizing selection of processing parameters of millgrinding sic/al composites[D]. Harbin: Harbin Institute of Technology, 2011.

[4] Lawn B R, Evans A G, Marshall D B. Elastic/plastic indentation damage in ceramics: the median/radial crack system[J]. Journal of the American Ceramic Society, 1980, 63(9-10): 574-581.

[5] Songmei Yuan, Huitao Fan, Muhanmmad Amin, Chong Zhang, & Meng Guo. A cutting force prediction dynamic model for side milling of ceramic matrix composites C/SiC based on rotary ultrasonic machining[J]. Int J Adv Manuf Technol (2016) 86:37–48.

A Reliable Defect Detection Method for Patterned Wafer Image Using Convolutional Neural Networks with the Transfer Learning

Ji-Hee Lee[1] and Jee-Hyong Lee[2*]

[1]Department of Semiconductor Display Engineering, Sungkyunkwan University, South Korea

[2] Department of Computer Science and Engineering, Sungkyunkwan University, South Korea

[*]Corresponding author: john@skku.edu

Abstract. In the semiconductor manufacturing process, wafer inspection images have valuable information on defects and wafer yield. It is worthy to analysis these images and there are many algorithms for this. Whenever new product designs are introduced, the algorithms are used to detect defects. Most new product designs have new types of defects. Hence, it is important to update the algorithms to cope with new types of defects. To update the algorithms, engineers collect the data and adjust a parameter values, such as threshold, to detect defects. It is time-consuming to collect enough data and only a knowledgeable engineer can find an appropriate parameter value. However, there is always a lack of engineer resources and time in the manufacturing industry. Due to these reasons, many algorithms can't be used reliably and have disappeared. Therefore, we propose the advanced method to update an algorithm easily using a deep learning model. Deep learning models achieve high accuracy with less engineer knowledge than traditional algorithms. But it also needs a lot of data to train the model, so we apply an advanced method of updating with a small amount of data in a short time. This is called as transfer learning. Once we train a deep learning model with high accuracy, we can update the model with a small amount of data. Transfer learning uses the information gained during the training. As a result, we make a model that is easy to update with high accuracy. In the experiments, we obtained 99% accuracy with sufficient training data. To update the model for the new data, we did transfer learning and got 97% accuracy. It was lower, but we only used 10% of the training data and 5% of the training time. Also, the accuracy we have obtained is enough to be used for defect analysis. Our approach outperformed and executed faster than traditional algorithms.

1. Introduction

In the semiconductor industry, detecting defect is an important issue because it is directly connected to the productivity and the revenue. There are eight different major semiconductor manufacturing processes and those processes are repeated until the wafer becomes a product. As a result, a wafer goes through several tens of manufacturing processes which could make defects. In order to reduce the overall process time, the wafers are not inspected in every process except when the inspection doesn't increase the process time. It is hard to detect defects with a limited inspection data. In addition, whenever a new type of wafer pattern with a new type of defect is introduced, an algorithm update is inevitable

To update the algorithm, it is necessary to have enough data to analyze the defects and time to adjust the parameter values, such as threshold, to detect defects. However, collecting enough data is not easy. Although every wafer has their inspection data, these are not labelled data. They need to be labelled manually and it is very time consuming. In addition, they are highly imbalance data. Normal data is always many times more than the defect data. Imbalanced data generally leads a wrong result. Most of the stable manufacturing processes have the same problem. Adjusting the parameter values are also difficult. Only knowledgeable engineers can decide the parameter value to detect.

There are a number of studies to detect and classify defects. Shankar and Zhong [1] showed how to find defects and visualize them. Tsai and Yang [2], Zontak and Cohen [3] also described how to compare two images and detect defects. These approaches used the golden image to detect a defect using a colour difference. Thus, the golden image should be prepared whenever a wafer needs to be inspected. Pak et al. [4] explained how to automatically set the threshold that separates objects and a background based on the grey-scale distribution. This algorithm is not easy to be applied to a patterned wafer. One of the famous algorithms is die-to-die comparison which is used commonly [5]. To compare dies, it needs to find a die boundary and to set a threshold. It requires a lot of engineer effort and is vulnerable to an ambiguous boundary. In addition to these algorithms, some machine learning based approaches are used for the defect classification [6], [7], [8]. Those approaches achieve high accuracy, but they need a lot of data to train.

All of the above-mentioned studies require a lot of time and effort from the engineers. However, there are limited engineer resources and time in the manufacturing industry. To overcome these difficulties, we propose the advanced method to detect defects using a deep learning model. The deep learning models we use is based on the Convolutional Neural Network (CNN) such as VGG16 [9], VGG19 [9] and Inception v3 [10] which are specialized for the image processing and a feature extraction. These models achieve high accuracy and require less engineer knowledge to detect defects. One disadvantages of these models is that they require a lot of training data. This is not suitable for frequent updating of a new type of wafer pattern. So, we apply the transfer learning [11], [12] to update the models easily. The transfer learning is a method of retraining a trained model for additional data. Once we train the model with sufficient data for the accuracy at the first time, we can update the model with small additional data in a short time without losing accuracy. Thus, we don't need to collect much data for updating and are able to fast respond. This is a great benefit to the industry and these benefits enable us to obtain a reliable model by updating the model as needed.

This paper is organized as follows. Section 2 states the problem defines. We explain the method and model architecture we used in section 3. Next, section 4 describes experiments and section 5 provides the conclusion.

2. Proposed method

In order to improve the detection accuracy, we adopt the Convolutional Neural Network (CNN) that shows the best result for the image processing. CNN uses a convolution layer instead of a fully connected layer. This convolution layer extracts features from the image easily. All of the images we use vary in complex patterns and colours depending on the manufacturing process. So, we use CNN that is specialized in extracting the features. At the beginning of this experiment, we evaluated a simple CNN models that have 3 to 7 layers and we got only 50% accuracy. Therefore, we apply three more complicate models including VGG16 [9], VGG19 [9] and Inception v3 [10]. We evaluate these models in aspects to accuracy and re-trainability. VGG16 and VGG19 are a simpler model than Inception v3. Only difference between VGG16 and VGG19 is a layer depth. Inception v3 is wider and deeper but it has fewer parameter than VGG16 and VGG19.

After training using these models, we use the transfer learning method. The transfer learning is proper to update the models with small effort. Once we train the models to have high accuracy, we can update the models for additional data. If the updated model doesn't have sufficient accuracy, we can notice it early since the transfer learning performs fast. During the transfer learning, we freeze some blocks as shown in Figure 1 and 2. By freezing the blocks, we can reuse the knowledge from the trained model. We only train some unfrozen blocks with additional data. Most of the blocks share the knowledge and some blocks learn the special feature from additional data. In this paper we call this retraining process as a fine-tuning.

2.1. Network architecture

VGG16, VGG19, Inception v3 are already proved for the image classification problem. We use the ImageNet weight for the training. After these models, we attach a dropout layer, global average pooling layer [13] and a dense layer. The dropout layer makes the model generalize. The global average pooling layer reduces the number of parameters in the model prior to the classification. Even if we use images of different size, we can still use the same parameter size using this layer before the classification layer. The last dense layer prepares a data structure for classification. We use a binary classifier because we only judge whether defect exists or not. We depict the number of blocks to freeze for fine-tuning in Figure 1 and 2.

2.1.1. VGGNet. VGG16 and VGG19 are generally called VGGNet that introduced the 2014 ILSVRC. Depends on the layer count, it is named VGG16 or VGG19. The most powerful point of these models is simplicity. They use only 3 x 3 convolution layer and max pooling. These models just stack the layers and right most is the top. We evaluate the performance depends on the layer count using these two models. VGG19 has more three layers than VGG16. Figure 1 shows the model architecture.

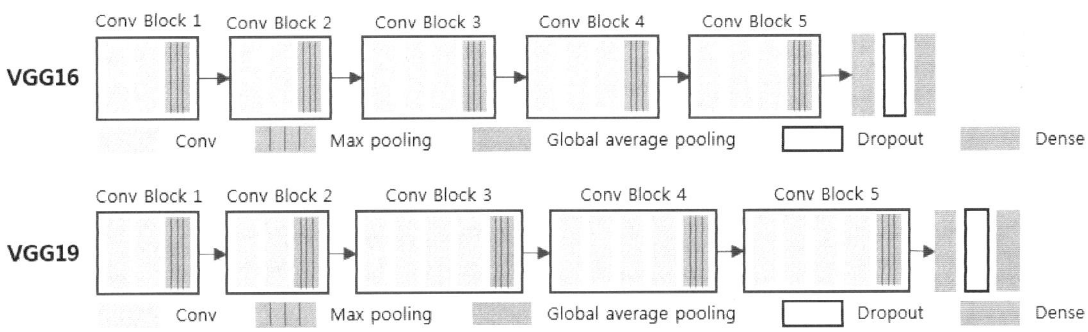

Figure 1. VGG16 and VGG19 model architecture

2.1.2. Inception v3. As seen in Figure 2, Inception v3 is a deeper and wider model than VGGNet. It is also introduced the 2014 ILSVRC and won. This model is composed of inception module that includes convolution layers and pooling layer in parallel. Inception module makes the model trainable even though deep and wide. We mark the inception module as a block. Despite of a wider and deeper model, it has fewer parameter than VGGNet. So, this model converges very quickly.

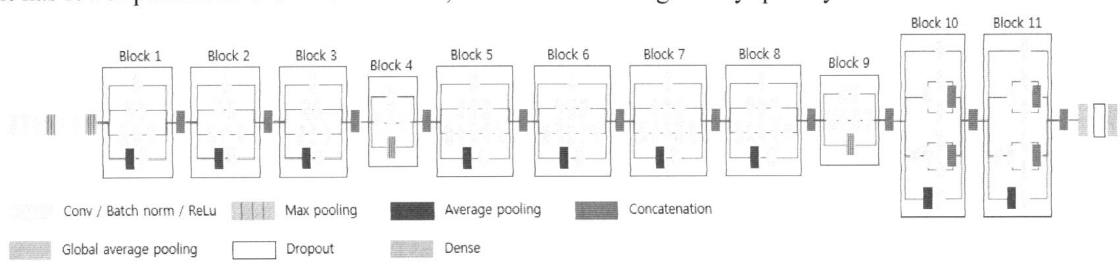

Figure 2. Inception v3 model architecture

2.2. Input transformation

The result image is 2048x2048 with a low resolution. This size is too large to train therefore we rescale the images to 512x512. We use the Keras library to train. This library provides several image transformation options. We apply four transformation including rescale, shear, zoom and horizontal flip. Table 1 shows each transformation setting type and value. These transformations are randomly applied to images when it loaded.

Table 1. Data transformation type

Transformation type	value
Rescale	0 ~ 1
Shear	0.2
Zoom	0.2
Horizontal Flip	True

3. Experiments

As mentioned above, we use three famous models to detect defects. Keras library we use provides the trained model using ImageNet dataset. So, we first fine-tune the provided models with the transfer learning for additional data. And then, we train the models with our training images. This is called the full training. During the full training, the models are taught the features of the wafer images. Finally, we fine-tune the trained models through the full training with transfer learning. We evaluate all these experiments in terms of accuracy and re-trainability. Building a reliable system requires the model that can be retrained quickly with high accuracy. We focus on this point during the experiments.

3.1. Dataset

The dataset we use is an image of inspection for CMP process from Samsung which is semiconductor company in South Korea. Once a wafer is processed, an inspection image is taken. We get these images and make a database. We only use the partial data generated in October 2018. The training data is ten types of wafer pattern. All types classified by the original method are highly imbalanced. We extract same count by the label to balance the data. After that, we modify some labels because the original method misclassifies some data. It makes a little imbalance. Table 2 provides a summary of the data. The error rate means the percentage of the misclassified label by the original method. Among ten types, we train and validate using seven types (A to G) of images. Remained three types (H to J) of images are used for fine-tuning and test. Aside from these images, we add additional images to check the performance of a deep learning model. Added additional images are not distinguished by the original method because of the ambiguous boundary or colour variation. We call these images as 'Unclassified'.

Table 2. Data description

Purpose	Type	Pass	Fail	Total	Error rate
Training &	A	267	361	628	7.50%
Validation	B	699	996	1695	11.30%
	C	486	586	1072	4.70%
	D	1863	2137	4000	5.90%
	E	295	391	686	7.00%
	F	461	703	1164	10.40%
	G	1873	2127	4000	3.40%
Test &	H	218	189	407	27.30%
Fine-tune	I	212	240	452	13.10%

	J	1522	1664	3186	7.30%
Test	Unclassified-A	150	150	300	-
	Unclassified-B	150	150	300	-
	Unclassified-C	150	150	300	-

3.2. Experimental setup

All experiments are performed on a 2.50GHz Intel Xeon with 256GB memory and NVDIA Quadro M6000 with 24GB Memory.

We use the pre-trained models provided from Keras library. It has already been proven that pre-trained models achieve better performance. The pre-trained models are trained using ImageNet [14]. ImageNet is large scale image database for deep learning.

3.3. Fine-tuning with pre-trained model

All models are pre-trained using ImageNet and already used for fine-tuning in other area. So, we fine-tune and then validate these models without any extra training. In order to fine-tune and validate the models, we use three test data (Type H, I, J). The total number of images we gather is approximately 4000, but we only use 1000 training and 500 validation samples to balance the type. Each type is included at the same rate. We train during 50 epochs with 5 batch size. We load the pre-trained models and freeze some bottom layers to fine-tune. We experiment on three conditions to evaluate how many layers are effective for fine-tuning. All training models are composed by the blocks, so we freeze using this block unit. We sequentially evaluate by freezing top two blocks, one block and only last fully-connected layer. Table 3 shows the accuracy and F1-score. VGGNet achieves around 87% accuracy but Inception v3 has maximum 70% accuracy. VGGNet is better than Inception v3 but all models are under 90%. These pre-trained models provided from Keras are trained by the large scale of images. However, those images include things that can be seen everywhere like cat and dog. They are different from the wafer image with repeated patterns. And even though a wafer pattern is same, the colour of image looks differently depends on process. Therefore, the frozen bottom blocks of three pre-trained models using ImageNet are more likely to have less relevant features with a wafer. Improper knowledge leads to low model accuracy.

Table 3. Fine-tuning result (Accuracy and F1-Score)

Unfrozen layer	VGG16		VGG19		Inception v3	
	Accuracy	F1 Score	Accuracy	F1 Score	Accuracy	F1 Score
Last layer	81.40%	0.82	81.00%	0.81	**70.00%**	0.7
1 block	87.00%	0.87	85.20%	0.85	63.00%	0.61
2 blocks	**87.40%**	0.87	**87.00%**	0.87	64.00%	0.62

3.4. Full training

To get more relevant bottom module with a wafer image, we train all models using seven types (A to G). Among the images, we randomly select 10,000 samples for training and 3,000 samples for validation. We train the models with 100 epochs and 5 batch size. All models achieve high accuracy. It is better performance than the original method which is about 93%. Figure 3 (a), Figure 3 (b) show the convergence speed is little different. Inception v3 is the fastest and VGG19 followed. The complicate model is generally slower than the simple model and it is harder to train. However, Inception v3 has high accuracy, fast convergence speed and short training time. Table 4 explains the full training result and shows that inception v3 is the best in all aspects. In addition, the training time of VGG16 and VGG19 shows if the model is same, the deeper model needs more time.

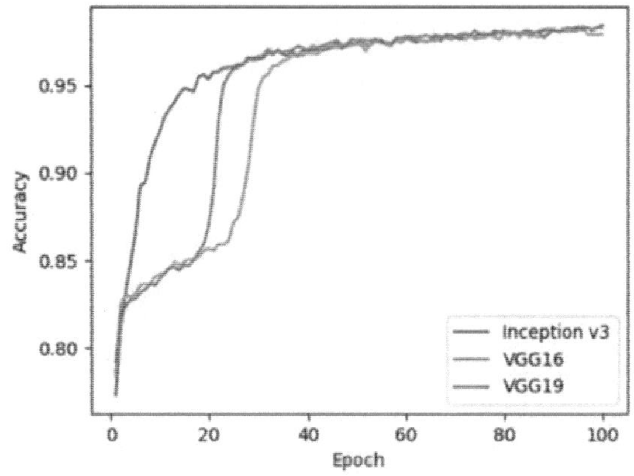

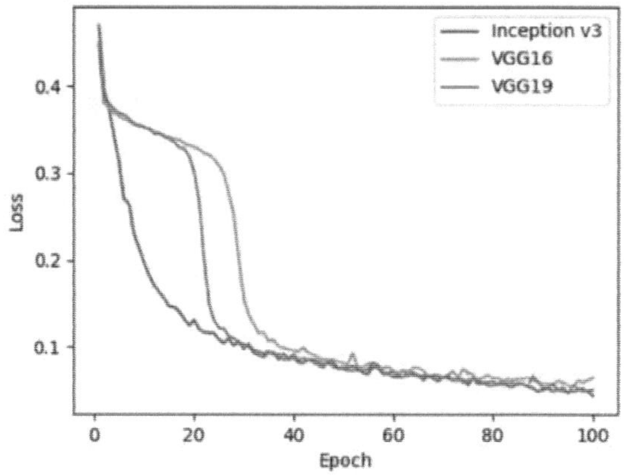

Figure 3. (a) Accuracy (b) Loss trend depends on models

Table 4. Full training results using data A to G in the dataset.

Model	Training time (hour)	Training		Validation		
		Accuracy	Loss	Accuracy	Loss	F1-Score
VGG16	41.2	98%	0.064	99%	0.038	0.99
VGG19	46.6	98%	0.050	99%	0.033	0.99
Inception v3	**36**	**99%**	**0.036**	**99%**	**0.025**	**0.99**

In Table 5, we evaluate two data types. One is 'Unclassified' data we mentioned, and the other is additional data.

Table 5. Evaluation Result using data H to J and 'Unclassified' in the dataset.

Data Type	VGG16		VGG19		Inception v3		Original method
	Accuracy	F1 Score	Accuracy	F1 Score	Accuracy	F1 Score	Accuracy
Unclassified	90%	0.90	91%	0.91	91%	0.91	Can't evaluate
Test (H, I, J)	73%	0.71	68%	0.65	80%	0.8	84%

'Unclassified' data that accounts for 25% of the total. Original method can't detect the defects but all models we trained get about 90% accuracy. This accuracy is not bad but a little lower than the training result. That's because these models don't have information about the ambiguous boundary.

In the industry area, reliability is as important as high accuracy. Therefore, we evaluate the additional data. The additional data are similar to the training data in terms of wafer images. However, the result is not satisfactory. It is worse than original method and also previous fine-tuning result. Even though the bottom module is similar, final high features like pattern are different from training data. It may lead to the poor result.

Through this experiment, we find that the deep learning models are appropriate to detect defects in terms of accuracy. But in other respects, it's a little different. Although 'Unclassified' data achieves the accuracy about 90%, additional data are not close to 90%. Both evaluation results show that only full training can't cover the new type of data. In addition, training time needs to be considered. In this experiment, the shortest time of training is 36 hours. In the manufacturing industry it is not acceptable that the model training takes more than one day since it means defective wafers are produced during one day without engineer's notice. This one-day training time doesn't even include the pre-processing time for the new training dataset.

3.5. Transfer Learning with full training weights

According to the previous experiments, the full training is also not enough to make a reliable detecting system. We need a training for additional data as short as possible. The transfer learning, we mentioned section 2, is useful to train for small amount of data in a short time. Based on the previous training result, we train additional data with freezing bottom blocks. The previous training weights contain useful low features in the bottom module. These features help the models get better accuracy. This experiment is done in the same way and same condition as section 3.3.

All models spend about 2 hours that is 5% of the previous training time and used 10% of the full training data. Table 6 shows the transfer learning result. VGG16 and VGG19 are achieved maximum 97% accuracy however Inception v3 is just 90%. All models are the best performance when training the models with unfrozen top two blocks. We also evaluate the 'Unclassified' in best experimental conditions. An original method can't evaluate at all, but we can detect defects using our method even when transfer learning. Table 7 shows the evaluation result. It is not bad accuracy considering that it is not trained data.

Figure 4 shows the accuracy and loss chart. We only draw the best performance in a chart. Contrary to the full training result, Figure 4 show that VGG19 converges fast in this experiment. When only the last layer is frozen, all models show noticeably low performance. We add three blocks condition in this experiment. Using this condition, we check the effect of the increase in the number of the frozen layers. When too many blocks are frozen, models can't train the high features to be used to detect. Less freeze the blocks, it may cause over-fitting. It also leads to poor performance.

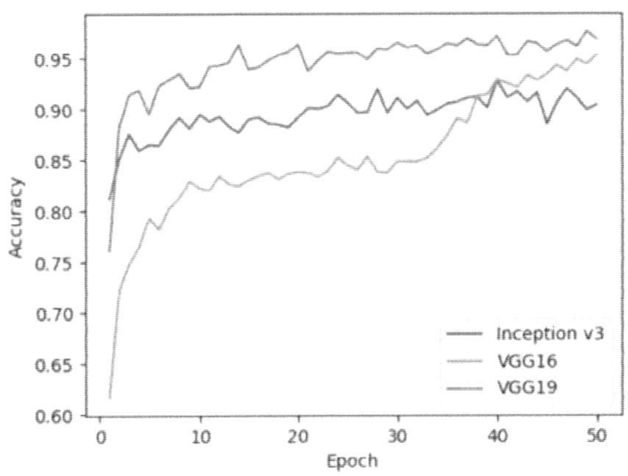

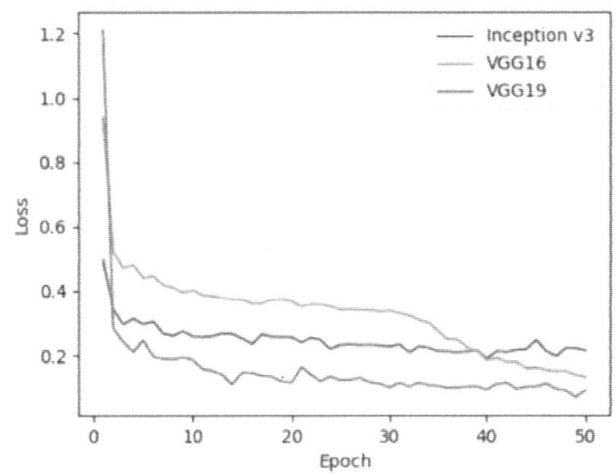

Figure 4. (a) Accuracy (b) Loss trend depending on models

Table 6. Transfer learning result using data H to J.

Unfrozen layer	VGG16		VGG19		Inception v3	
	Accuracy	F1 Score	Accuracy	F1 Score	Accuracy	F1 Score
Last layer	57.60%	0.49	67.60%	0.65	83.80%	0.84
1 block	96.80%	0.97	92.60%	0.93	87.60%	0.88
2 blocks	**97.00%**	0.97	**97.00%**	0.97	**90.00%**	0.9
3 blocks	87.00%	0.87	87.00%	0.87	65.00%	0.6

Table 7. Transfer learning evaluation using 'Unclassified' data.

Unfrozen layer	VGG16		VGG19		Inception v3	
	Accuracy	F1 Score	Accuracy	F1 Score	Accuracy	F1 Score
2 blocks	**90.22%**	0.90	**86.11%**	0.86	89.22%	0.89

Through this experiment, we find that transfer learning is suitable for fast respond. Once we train the models with high accuracy, we can update the models with small amount of data in a short time. It is slightly lower than the full training, but it enables to analyze the defects fast. This makes the models reliable.

4. Conclusion

In the manufacturing industry, there are limited engineer resources and time. With this limitation, we detect defects whenever a new type of wafer pattern is introduced. To cope with this problem, we propose the advanced method to update using a deep learning model. Deep learning models we use are specialized for the image processing. It leads to high accuracy models to detect defects. And we can detect defects that can't be detected by an original method. However, these models need a lot of data for a training. So, we apply the method to update with small amount of data in a short time. This is called transfer learning. This reduces the burden to collect data. It is great beneficial in the industry area. Using our proposal, we achieve 99% accuracy in the full training. Then, we obtained 97% accuracy when transfer learning for a new type of data. This is slightly lower than the full training, but it is sufficient for fast respond to analyze the defects. This approach enables us to obtain a reliable model by updating the model as needed.

As the process and the design pattern become more complicated, it is important to not only judge whether defects exist but also classify defects. Therefore, this approach needs to be improved to classify the detailed types of defects. We remain this as a future work.

References

[1] NG Shankar, ZW Zhong. Microelectronic engineering **77**: p. 337-346 (2005)
[2] DM Tsai and RH Yang. Image and Vision Computing **23**: p. 1094-1101 (2005)
[3] M Zontak and I Cohen. Machine Vision and Applications **21**: 129-141 (2010)
[4] M Pak, J Park and S Kim. Computer Science and its Applications. Springer, Berlin, Heidelberg. p. 1085-1091 (2015)
[5] I Isomura. U.S. Patent No. 7,630,535. (2009)
[6] KB Lee, S Cheon and CO Kim. IEEE Transactions on Semiconductor Manufacturing **30**: p. 135-142 (2017)
[7] T Nakazawa and DV Kulkarni. IEEE Transactions on Semiconductor Manufacturing **31**: p. 309-314 (2018)
[8] G Tello, OY Al-Jarrah, PD Yoo, Y Al-Hammadi, S Muhaidat, and U Lee. IEEE Transactions on Semiconductor Manufacturing **31**: p. 315-322 (2018)
[9] K Simonyan and A Zisserman. arXiv preprint arXiv:1409.1556 (2014).
[10] C Szegedy, W Liu, Y Jia, P Sermanet, S Reed, D Anguelov, D Erhan, V Vanhoucke and A Rabinovich.IEEE. computer vision and pattern recognition (2015)
[11] SJ Pan and Q Yang. IEEE Transactions on knowledge and data engineering **22**: p. 1345-1359 (2010)
[12] L Shao, F Zhu and X Li. IEEE transactions on neural networks and learning systems **26**: p. 1019-1034 (2015)
[13] M Lin, Q Chen and S Yan *arXiv preprint arXiv:1312.4400* (2013)
[14] J Deng, W Dong, R Socher, LJ Li, K Li and L Fei-Fei. "Imagenet: A large-scale hierarchical image database.": p. 248-255 (2009)

Analysis of Allowable Assembly Forces for Composite Laminates

Tingyu Zhang[1,a], Chong Zhang[1], Jianshe Zhao[1] and Anru Guo[1]

[1]Institute of Aerospace Materials and Technology,South 1st Dahongmen Road, Fengtai District, Beijing, China, 100076

[a]tingyubeijing@163.com

Abstract. Composite structures have been widely used in aerospace field due to their low density, high strength, and high stiffness to weight ratio. In the aerospace field, composite laminates are widely used in the skin of the airplane. Due to the complexity of the composite molding process, and the difficulty of machining composite laminates, the manufacturing precision of the skin is usually not very high. Hence the assembly force is applied to the skin to ensure the composite laminates fit the framework in the assembly process of an airplane. However, too large assembly force leads to the assembly out of tolerance owning to the over deformation and leads to the stress of the skin out of tolerance. Therefore, this paper proposes a method of analyzing the allowable assembly force. With the help of mechanics and physics, the mechanical relationship between the assembly forces and the deformation of a skin is modeled. On the basis of the model and the experimental date of a skin assembly process, the mathematical relationship between the assembly forces and the deformation of the skin is established. The allowable assembly forces are determined based on the mathematical relationship and the allowable deformation. This paper provides an optimization methodology of assembly force optimization.

1. Introduction

Composite structures have been widely used in practice due to their low density, high strength, and high stiffness-to-weight ratio [1]. In aerospace field, composite laminates are used in the skin of the airplane. Due to the complexity of the composite molding process, and the difficulty of machining composite laminates, the manufacturing precision of the skin is usually not very high. Hence the assembly force is applied to the skin to ensure the composite laminates fit the framework in the assembly process of an airplane. However, too large assembly force leads to the assembly out of tolerance owning to the over deformation and leads to the stress of the skin out of tolerance.

In recent years, research has been done in the area of assembly deviation modeling and analysis for the assembly of sheet metal compliant parts. Liu and Hu [2] proposed an offset element method to predict assembly deviation of sheet metal for a one dimensional model by combining engineering structural mechanics with statistical methods. Moreover, assembly tolerances in series and in parallel were evaluated by using linear mechanics [3], which showed that the assembly deviation can be less than the "stacked up" of component deviations in compliant part assembly due to the fact that assembly deviation was affected not only by the geometry of the components but also by the stiffness of the components in parallel assembly. Liu and Hu [4] expanded the model to compliant sheet metal parts with 2D or 3D free-form surfaces by using FEM to construct a sensitivity matrix relating the incoming part deviation to the output assembly deviation. This method was called the method of influence coefficient (MIC). MIC was further extended to multi-station assembly processes with compliant parts by Camelio et al. [5] following the concepts of the stream-of-variation theory [6]. It is assumed that sources of variation are independent in the paper [5]. However, the independence assumption may not always be adequate [7]. Camelio et al. [7] presented the effect of geometric covariance of sources of variation on the calculation of assembly variation of compliant parts with consideration of the dependence of sources of variation. All the papers above are about the assembly variation.

The purpose of this paper is to develop a method of determining the allowable force. First of all, the mechanical relationship between the assembly forces and the deformation is modeled, on the basis of the mechanics and physics. And then, an experiment of a skin is done to obtain the assembly force and the deformation of the skin. In further, the mathematical relationship between the assembly force and the deformation of the skin is constructed. Finally, the allowable assembly force is determined based on the allowable deformation and the mathematical relationship.

2. Derivation of the Relationship between the Assembly Force and the Assembly Deformation

Please follow these instructions as carefully as possible so all articles within a conference have the same style to the title page. This paragraph follows a section title so it should not be indented.

A lot of papers study how to model the assembly deviation for the sheet metal assembly [2-7]. For an isotropic sheet metal assembly, the relationship between the assembly force in the z direction and the structure deformation in the z direction for isotropic sheet metal is represented in Eq. 1, which was developed in the MIC literature [5, 7].

$$\mathbf{F}=\mathbf{K}\times\mathbf{V} \tag{1}$$

where matrix F is the force matrix, matrix V the deformation matrix, matrix K the stiffness matrix. Matrix K represents the force exerted to the sheet metal if the sheet metal produces unit deformation.

It is noticed that the relationships between the assembly force and the structure deformation for isotropic sheet metal and anisotropic composite laminates are different. Because composite material is anisotropic material, coupling effects exist in non-symmetric composite laminates. For example, tension (or compression) deformation and shear deformation will be created when bending assembly load is exerted on a non-symmetric composite laminate [8].

To model dimensional deviation for the assembly process of composite laminates, a new finite element equation expressing the relationship between the external forces and the deformation, as shown in Eq.2, is derived on the basis of finite element theory. The derivation process is presented in [9].

$$\begin{bmatrix} \mathbf{K}^{uu} & \mathbf{K}^{uv} & \mathbf{K}^{u\omega} \\ \mathbf{K}^{vu} & \mathbf{K}^{vv} & \mathbf{K}^{v\omega} \\ \mathbf{K}^{wu} & \mathbf{K}^{\omega v} & \mathbf{K}^{\omega\omega} \end{bmatrix}_{3n\times 3n} \times \begin{Bmatrix} \mathbf{u} \\ \mathbf{v} \\ \boldsymbol{\omega} \end{Bmatrix}_{3n\times 1} = \begin{Bmatrix} \mathbf{F}^{u} \\ \mathbf{F}^{v} \\ \mathbf{F}^{\omega} \end{Bmatrix}_{3n\times 1} \tag{2}$$

where each sub-matrix in Eq. 2 can be further represented as

$$\mathbf{K}^{rs} = \begin{bmatrix} k_{11}^{rs} & k_{11}^{rs} & \Lambda & k_{1n}^{rs} \\ k_{21}^{rs} & k_{22}^{rs} & \Lambda & k_{2n}^{rs} \\ \mathrm{M} & \mathrm{M} & \mathrm{O} & \mathrm{M} \\ k_{n1}^{rs} & k_{n2}^{rs} & \Lambda & k_{nn}^{rs} \end{bmatrix} \quad (r,\, s = u, v, \omega) \tag{3}$$

$$\mathbf{r} = \begin{bmatrix} r_1 & r_2 & \Lambda & r_n \end{bmatrix}^T \quad (\mathbf{r} = \mathbf{u},\, \mathbf{v},\, \boldsymbol{\omega}) \tag{4}$$

$$\mathbf{F}^r = \begin{bmatrix} F_1^r & F_2^r & \Lambda & F_n^r \end{bmatrix}^T \quad (r = u, v, \omega) \tag{5}$$

here the mechanics meaning of the term $\mathbf{K}_{ij}^{rs}$ in the structure stiffness matrix is the force in the r direction needed from the ith node to produce the unit displacement in the s direction at the jth node, and n is the total number of nodes.

In the assembly process, there are only forces in the z direction in the assembly process. It indicates that

$$\mathbf{F}^u = \mathbf{0}_{n\times 1} \text{ and } \mathbf{F}^v = \mathbf{0}_{n\times 1} \tag{6}$$

Substitute Eq. 6 into Eq. 2, and with proper derivations, we get

$$\mathbf{K}^{\omega} \times \boldsymbol{\omega} = \mathbf{F}^{\omega} \tag{7}$$

where K^{ω} is called as the equivalent structure stiffness matrix to distinguish it from the structure stiffness matrix K. The expression of K^{ω} can be presented as

$$\begin{aligned} \mathbf{K}^{\omega} = {} & \mathbf{K}^{\omega\omega} - \mathbf{K}^{\omega u} \times (\mathbf{K}^{vu} - \mathbf{K}^{v\omega} \times (\mathbf{K}^{u\omega})^{-1} \times \mathbf{K}^{uu})^{-1} \times (\mathbf{K}^{v\omega} \times (\mathbf{K}^{u\omega})^{-1} \times \mathbf{K}^{uv} - \mathbf{K}^{vv}) \\ & \times (\mathbf{K}^{uu} \times (\mathbf{K}^{vu} - \mathbf{K}^{v\omega} \times (\mathbf{K}^{u\omega})^{-1} \times \mathbf{K}^{uu})^{-1} \times (\mathbf{K}^{v\omega} \times (\mathbf{K}^{u\omega})^{-1} \times \mathbf{K}^{uv} - \mathbf{K}^{vv}) + \mathbf{K}^{uv})^{-1} \times \mathbf{K}^{u\omega} \\ & - \mathbf{K}^{\omega v} \times (\mathbf{K}^{uu} \times (\mathbf{K}^{vu} - \mathbf{K}^{v\omega} \times (\mathbf{K}^{u\omega})^{-1} \times \mathbf{K}^{uu})^{-1} \times (\mathbf{K}^{v\omega} \times (\mathbf{K}^{u\omega})^{-1} \times \mathbf{K}^{uv} - \mathbf{K}^{vv}) + \mathbf{K}^{uv})^{-1} \times \mathbf{K}^{u\omega} \end{aligned} \tag{8}$$

Eq.7 indicates the mechanical relationship between the force in the z direction and the deformation of the assembly in the z direction.

3. Allowable Assembly Force of a skin of an airplane

The gap always exists between the framework of an airplane and the skin of an airplane due to the forming error of the skin and the machining error and the assembling error of the framework. To close the gap, the assembly force has to been applied to the skin. However, too large assembly forces can lead to the over deformation of a skin. Therefore, it is important to obtain the allowable assembly force. Since the allowable assembly force is the extreme value, it is harmful to a skin if the allowable assembly force is attained only by the experiment. Thereby, some experiments are done to monitor the values of the assembly forces and the assembly deformation which are much smaller than the extreme values. Substitute the values of the assembly forces and the assembly deformation to Eq. 7 to construct the mathematical relationship of the assembly force and the assembly deformation of a skin. On the basis of the mathematical relationship and the allowable deformation of a skin, the allowable assembly force can be obtained.

3.1. Measuring of the Assembly Force and the Assembly Deformation of a Skin.

To measure the assembly force and assembly deformation of a skin, an measuring equipment is designed and manufactured. As shown in Figure 1, the measuring equipment consists of three parts: ① base, ②pressure mechanism, and ③deformation measuring mechanism. The base is used to fix the skin of an airplane. The pressure mechanism can slide along the rail to measure the pressure applied to any region of the skin through the manometers, and the deformation measuring mechanism can also slide along the rail to measure the deformation of any region of the skin.

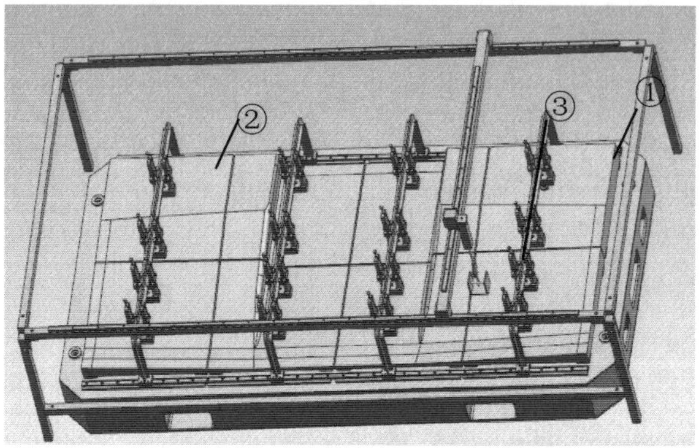

Figure 1 Measuring equipment

Four four pressure points and the four deformation measurement points are selected. The measurement results of the assembly force and the assembly deformation are listed in Table 1.

Table 1. Measurement results of the assembly force and the assembly deformation

Pressure points	Assembly force（N）	Deformation（mm）			
		a	b	c	d
1	45	2.47	1.8	1.51	1.5
2	48.8	2.57	3.01	2.64	2.35
3	55.3	2.28	2.07	2.26	2.16
4	50	1.43	1.79	1.98	3.1

3.2. Construction of the Relationship of the Assembly Force and the Assembly Deformation

In this case, there are four pressure points and four deformation measurement points. Therefore, Eq.7 can be further represented as

$$\begin{bmatrix} k_{11}^{\omega\omega} & k_{12}^{\omega\omega} & k_{13}^{\omega\omega} & k_{14}^{\omega\omega} \\ k_{21}^{\omega\omega} & k_{22}^{\omega\omega} & k_{23}^{\omega\omega} & k_{24}^{\omega\omega} \\ k_{31}^{\omega\omega} & k_{32}^{\omega\omega} & k_{33}^{\omega\omega} & k_{34}^{\omega\omega} \\ k_{41}^{\omega\omega} & k_{42}^{\omega\omega} & k_{43}^{\omega\omega} & k_{44}^{\omega\omega} \end{bmatrix} \times \begin{bmatrix} \omega_1 \\ \omega_2 \\ \omega_3 \\ \omega_4 \end{bmatrix} = \begin{bmatrix} F_1^{\omega} \\ F_2^{\omega} \\ F_3^{\omega} \\ F_4^{\omega} \end{bmatrix} \tag{9}$$

Based on the measurement results shown in Table 1, Eq.9 can be represented as

$$\begin{gathered} \begin{bmatrix} k_{11}^{\omega\omega} & k_{12}^{\omega\omega} & k_{13}^{\omega\omega} & k_{14}^{\omega\omega} \\ k_{21}^{\omega\omega} & k_{22}^{\omega\omega} & k_{23}^{\omega\omega} & k_{24}^{\omega\omega} \\ k_{31}^{\omega\omega} & k_{32}^{\omega\omega} & k_{33}^{\omega\omega} & k_{34}^{\omega\omega} \\ k_{41}^{\omega\omega} & k_{42}^{\omega\omega} & k_{43}^{\omega\omega} & k_{44}^{\omega\omega} \end{bmatrix} \times \begin{bmatrix} 2.47 \\ 1.8 \\ 1.51 \\ 1.5 \end{bmatrix} = \begin{bmatrix} 45 \\ 0 \\ 0 \\ 0 \end{bmatrix}, \begin{bmatrix} k_{11}^{\omega\omega} & k_{12}^{\omega\omega} & k_{13}^{\omega\omega} & k_{14}^{\omega\omega} \\ k_{21}^{\omega\omega} & k_{22}^{\omega\omega} & k_{23}^{\omega\omega} & k_{24}^{\omega\omega} \\ k_{31}^{\omega\omega} & k_{32}^{\omega\omega} & k_{33}^{\omega\omega} & k_{34}^{\omega\omega} \\ k_{41}^{\omega\omega} & k_{42}^{\omega\omega} & k_{43}^{\omega\omega} & k_{44}^{\omega\omega} \end{bmatrix} \times \begin{bmatrix} 2.57 \\ 3.01 \\ 2.64 \\ 2.35 \end{bmatrix} = \begin{bmatrix} 0 \\ 48.8 \\ 0 \\ 0 \end{bmatrix}, \\ \begin{bmatrix} k_{11}^{\omega\omega} & k_{12}^{\omega\omega} & k_{13}^{\omega\omega} & k_{14}^{\omega\omega} \\ k_{21}^{\omega\omega} & k_{22}^{\omega\omega} & k_{23}^{\omega\omega} & k_{24}^{\omega\omega} \\ k_{31}^{\omega\omega} & k_{32}^{\omega\omega} & k_{33}^{\omega\omega} & k_{34}^{\omega\omega} \\ k_{41}^{\omega\omega} & k_{42}^{\omega\omega} & k_{43}^{\omega\omega} & k_{44}^{\omega\omega} \end{bmatrix} \times \begin{bmatrix} 2.28 \\ 2.07 \\ 2.26 \\ 2.16 \end{bmatrix} = \begin{bmatrix} 0 \\ 0 \\ 55.3 \\ 0 \end{bmatrix}, \begin{bmatrix} k_{11}^{\omega\omega} & k_{12}^{\omega\omega} & k_{13}^{\omega\omega} & k_{14}^{\omega\omega} \\ k_{21}^{\omega\omega} & k_{22}^{\omega\omega} & k_{23}^{\omega\omega} & k_{24}^{\omega\omega} \\ k_{31}^{\omega\omega} & k_{32}^{\omega\omega} & k_{33}^{\omega\omega} & k_{34}^{\omega\omega} \\ k_{41}^{\omega\omega} & k_{42}^{\omega\omega} & k_{43}^{\omega\omega} & k_{44}^{\omega\omega} \end{bmatrix} \times \begin{bmatrix} 1.43 \\ 1.79 \\ 1.98 \\ 3.1 \end{bmatrix} = \begin{bmatrix} 0 \\ 0 \\ 0 \\ 50 \end{bmatrix} \end{gathered} \tag{10}$$

Based on Eq.10, the stiffness matrix $\mathbf{K}^{\omega\omega}$ can be obtained as

$$\mathbf{K}^{\omega\omega} = \begin{bmatrix} 2.0033 & 1.4599 & 1.2247 & 1.2166 \\ 1.1135 & 1.3042 & 1.1439 & 1.0182 \\ 1.6369 & 1.4861 & 1.6225 & 1.5507 \\ 0.9518 & 1.1915 & 1.3179 & 2.0634 \end{bmatrix}$$

3.3 Calculation of the Allowable Assembly Force

Since the allowable deformation of the skin is 5mm, the allowable assembly force is obtained as

$$\begin{bmatrix} F_1^{\omega} \\ F_2^{\omega} \\ F_3^{\omega} \\ F_4^{\omega} \end{bmatrix} = \begin{bmatrix} k_{11}^{\omega\omega} & k_{12}^{\omega\omega} & k_{13}^{\omega\omega} & k_{14}^{\omega\omega} \\ k_{21}^{\omega\omega} & k_{22}^{\omega\omega} & k_{23}^{\omega\omega} & k_{24}^{\omega\omega} \\ k_{31}^{\omega\omega} & k_{32}^{\omega\omega} & k_{33}^{\omega\omega} & k_{34}^{\omega\omega} \\ k_{41}^{\omega\omega} & k_{42}^{\omega\omega} & k_{43}^{\omega\omega} & k_{44}^{\omega\omega} \end{bmatrix} \times \begin{bmatrix} 5 \\ 5 \\ 5 \\ 5 \end{bmatrix} = \begin{bmatrix} 29.522 \\ 22.899 \\ 31.4816 \\ 27.6234 \end{bmatrix} \tag{11}$$

Therefore, the allowable assembly force of the skin is around 29N at point 1, around 23 N at point 2, around 31N at point 3, and around 28 N at point 4.

4. Conclusions

This paper developed a method of determining the allowable assembly forces and optimizing the assembly forces. In the method, an equivalent stiffness matrix is defined to construct the mechanical relationship between the assembly forces and structure deformation with the help of mechanics and physics. One case study of a skin assembly process is conducted. Depending on the experimental data of the assembly force and skin deformation and the mechanical relationship, the mathematical relationship between the assembly forces and the skin deformation is modeled. The allowable assembly forces are finally determined based on the allowable skin deformation.

References

[1] Mallick PK 1993 Fiber-Reinforced Composites: Materials, Manufacturing, and Design, Marcel Dekker, New York.

[2] Liu S and Hu S 1995 An Offset Element Model and Its Applications in Predicting Sheet Metal Assembly Variation *Int. J. Mach. Tools Manuf.* **35**(11) p 1545–57.

[3] Liu S, Hu S and Woo T 1996 Tolerance Analysis for Sheet Metal Assemblies *ASME J. Mech. Des.*, **118**(1) p 62–67.

[4] Liu S and Hu S 1997 Variation Simulation for Deformable Sheet Metal Assemblies Using Finite Element Methods *ASME J. Manuf. Sci. Eng.* **119**(3) p 368–374.

[5] Camelio J, Hu S and Ceglarek D 2003 Modeling Variation Propagation of Multi-Station Assembly System With Compliant Parts *ASME J. Mech.Des.*, **125**(4) p 673–681.

[6] Jin J and Shi J 1999 State Space Modeling of Sheet Metal Assembly for Dimensional Control *ASME J. Manuf. Sci. Eng.*, **121**(4) p 756–762.

[7] Camelio J, Hu S and Marin S 2004 Compliant Assembly Variation Analysis Using Component Geometric Covariance *ASME J. Manuf. Sci.Eng.* **126**(2) p 355–360.

[8] Nettles A 1994 Basic Mechanics of Laminated Composite Plates *NASA Reference Publication* **1351** p 33–40 and p 63–72.

[9] Zhang T and Shi J 2016 Stream of Variation Modeling and Analysis for Compliant Composite Part Assembly—Part I: Single-Station Processes *ASME J. Manuf. Sci.Eng.* **138** p 121003-1-121003-15.

IOP Conf. Series: Materials Science and Engineering **647** (2019) 012012 doi:10.1088/1757-899X/647/1/012012

Effects of Process Parameters on Cutting Force and Tool Temperature in Drilling Based on Finite Element Simulation

Yuan Gao*, Jie Wang, Qi Wang, Xin Li, Tingyu Zhang and Yang Zhong

Aerospace research institute of materials & processing technology, Beijing, China

* E-mail: zzgaoyuan@163.com

Abstract. Cutting force and tool temperature play significant roles in the drilling process because they are strongly related to drilling performance such as surface roughness, tool wear and tool breakage etc. For difficult-to-machine materials, like titanium alloys, the drilling operation is considered to be complex due to their mechanical and thermal properties, and it is important to find suitable process parameters to avoid too large cutting force or too high tool temperature, which may deteriorate the drilling performance. This paper is dedicated to studying the effects of process parameters, including spindle speed, feed rate and helix angle, on cutting force and tool temperature in drilling based on finite element simulation. A series of drilling simulations are carried out to generate the simulated cutting force and tool temperature data under different drilling conditions. By data fitting, the steady cutting force and tool temperature are predicted. Based on the Taguchi method, the effects of process parameters on steady cutting force and tool temperature are analysed. This paper can provide a basis for suitable selection of process parameters in drilling.

1. Introduction

Titanium alloys are commonly used in the recent years, especially in aerospace and biomedical industries [1] due to their exceptional strength to weight ratio, high temperature performance and corrosion resistance [2]. However, these alloys belong to difficult-to-machine materials because of their low thermal conductivity, high chemical reactivity and low modulus of elasticity [3], leading to large cutting forces and high tool temperatures during machining. Cutting force and tool temperature play important roles because they are strongly related to machining performance such as surface roughness, tool wear and tool breakage [4] etc. It is of great significance to find suitable process parameters before machining to avoid too large cutting force or too high tool temperature during machining, which may deteriorate the final machining quality.

Machining simulation by finite element method can provide an effective way to study the cutting force and tool temperature during machining. Tuğrul and Altan [5] developed a methodology for simulating the cutting process in flat end milling operation. Chip flow, cutting forces, tool stresses and temperatures are predicted by using finite element analysis. Sulaiman et al. [6] simulated the cutting force and tool temperature during high speed machining of AISI 4340 steel, and found the optimum tool rake angle that resulted in the smallest cutting force and the lowest tool temperature. Muhammad et al. [7] developed a three-dimensional simulation model of drilling process and investigated the effect of an external heat source on thrust force and torque, demonstrating that hot drilling had advantages in reducing thrust force and torque.

This paper is dedicated to simulating the cutting force and tool temperature in the drilling operations, and analysing the effects of process parameters, including spindle speed, feed rate and helix angle, on cutting force and tool temperature. The paper is organized as follows: In section 2, the layout of the drilling simulation and the definitions of several process parameters are demonstrated in details. The instantaneous cutting force and tool temperature data can be generated by simulations, and by data fitting, the steady cutting force and tool temperature can be predicted. In section 3, a series of drilling simulations are carried out under different drilling conditions, and the effects of process parameters on steady cutting force and tool temperature are analysed based on the Taguchi method. Finally, conclusions of this study are drawn in section 4.

2. Description of simulation model for drilling

Figure 1 shows the layout of the simulation model for drilling. Cuboid workpiece is adopted in the simulation, with the length L_w, the width W_w and the height H_w of the workpiece equalling to 12 mm, 12 mm and 10 mm, respectively. The workpiece material is defined as Ti6Al4V, while the drill material is defined as cemented carbide. A standard drill with diameter D_0 equalling to 6mm is used. The helix angle H_a usually varies from 18 degrees to 30 degrees, and is to be investigated in this study. The other relevant tool geometrical parameters are given as follows: the body diameter clearance is equal to 0.15 mm; the web thickness is equal to 1 mm; the flute radius is equal to 1.45 mm; the flute length is equal to 5 mm; and the edge radius is equal to 0.04 mm.

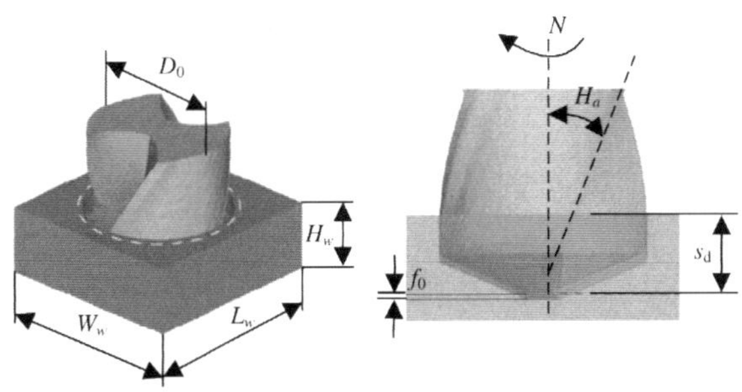

Figure 1. Layout of the simulation model for drilling.

The cutting parameters, including the spindle speed N and the feed rate f_0, are to be investigated in this study. The starting depth s_d is defined as the embedded depth of the drill into the workpiece at the beginning of the simulation time. In this study, the value of starting depth is set as 2.2 mm.

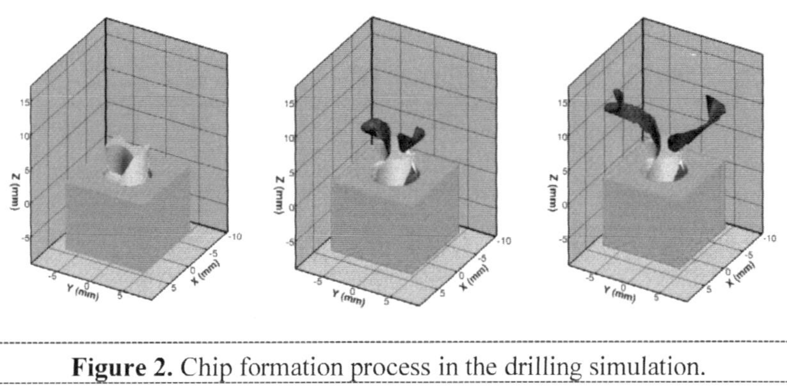

Figure 2. Chip formation process in the drilling simulation.

Figure 2 shows the chip formation process in the drilling simulation. The total angle of rotation of the drill is 1080 degrees. The simulated cutting force and tool temperature data are shown in Figure 3.

The cutting forces in the tangential direction and radial direction are theoretically equal to zero, while the cutting force in the axial direction is principal. In this study, only the axial cutting force is considered, and in the rest part of this article, the concept of cutting force represents the axial cutting force.

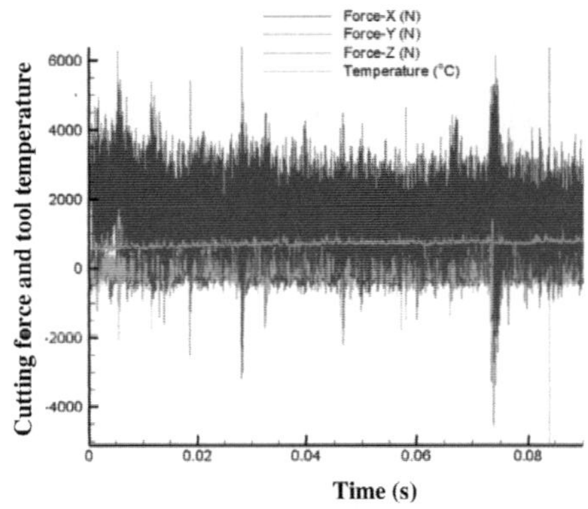

Figure 3. Simulated cutting forces and tool temperature data.

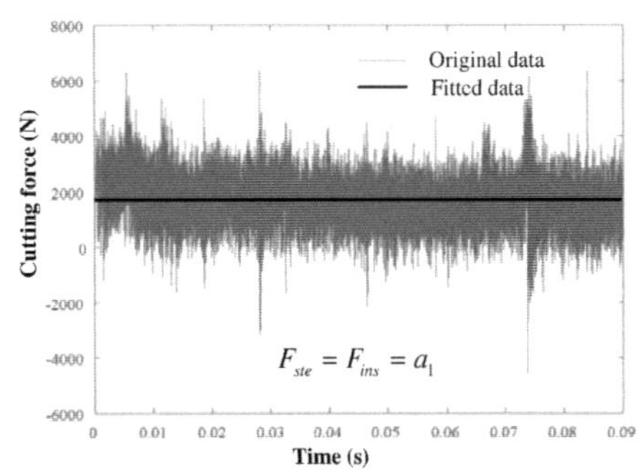

Figure 4. Data fitting for cutting force.

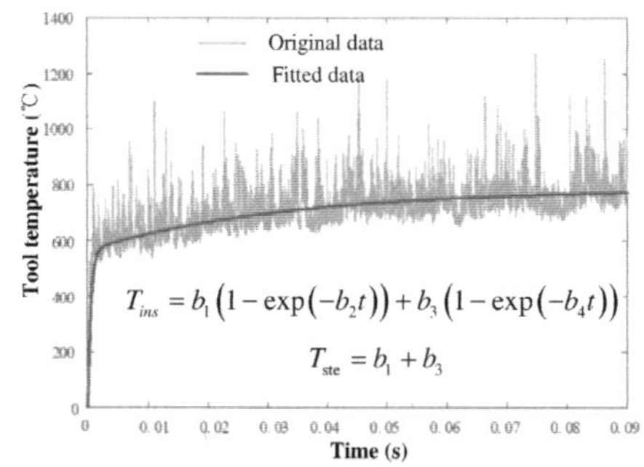

Figure 5. Data fitting for tool temperature.

The instantaneous cutting force and tool temperature fluctuate severely during the simulation period, as shown in Figure 4 and Figure 5. To make it easier to compare, a set of data fitting is carried out for both cutting force and tool temperature. The instantaneous cutting force F_{ins} can be expressed as follows:

$$F_{ins} = a_1 \tag{1}$$

where a_1 is a constant fitted from the original cutting force data shown in Figure 4. Therefore, the cutting force in the steady state F_{ste} can also be predicted to be equal to this constant:

$$F_{ste} = a_1 \tag{2}$$

The simulated tool temperature increases from the initial value and will reach to a steady state with enough time. The instantaneous tool temperature T_{ins} can be expressed as follows:

$$T_{ins} = b_1\left(1-\exp\left(-b_2 t\right)\right) + b_3\left(1-\exp\left(-b_4 t\right)\right) \tag{3}$$

where b_1, b_2, b_3, b_4 are four constants fitted from the original tool temperature data shown in Figure 5. Therefore, the steady tool temperature T_{ste} can be achieved when time t is infinite, and can be expressed as follows:

$$T_{ste} = b_1 + b_3 \tag{4}$$

3. Effects of process parameters on cutting force and tool temperature

In this study, a series of drilling simulations are carried out with different combinations of process parameters, including spindle speed, feed rate and drill helix angle.

Table 1. Process parameter levels.

Symbol	Parameter	Unite	Parameter levels Level 1	Level 2	Level 3
A	N	rpm	1000	2000	3000
B	f_0	mm/rev	0.2	0.3	0.4
C	H_a	degree	18	24	30

A L27 orthogonal array with 3 columns and 27 rows is applied. Each column represents one process parameter, and each process parameter has three levels, as given in Table 1. After 27 set of drilling simulations, the fitting coefficients are obtained from the simulated data, and the steady cutting force and tool temperature are predicted according to Eq. 2 and Eq. 4. The fitting coefficients and the corresponding cutting force and tool temperature results are summarized in Table 2.

Table 2. Fitting coefficients, cutting force and tool temperature results

No.	Parameters N	f_0	H_a	Fitting coefficients a_1	b_1	b_2	b_3	b_4	Results F_{ste}	T_{ste}	S/N ratio (dB) F_{ste}	T_{ste}
01	1000	0.2	18	1474	140.4	14.58	397.5	744.5	1474	537.9	-63.37	-54.61
02	1000	0.2	24	1390	113.8	17.35	421.1	775.6	1390	534.9	-62.86	-54.57
03	1000	0.2	30	1328	123.2	18.72	415.6	883.6	1328	538.8	-62.46	-54.63
04	1000	0.3	18	1978	179.4	17.98	438.5	679.6	1978	617.9	-65.92	-55.82
05	1000	0.3	24	1831	170.6	22.1	445.8	742.9	1831	616.4	-65.25	-55.80
06	1000	0.3	30	1805	175	13.32	467.8	766.9	1805	642.8	-65.13	-56.16
07	1000	0.4	18	2356	212	16.92	463.6	630	2356	675.6	-67.44	-56.59
08	1000	0.4	24	2225	202.7	19.74	465.9	553.5	2225	668.6	-66.95	-56.50
09	1000	0.4	30	2158	191.9	30.19	448.7	581.1	2158	640.6	-66.68	-56.13
10	2000	0.2	18	1440	155.8	43.87	517.5	1545	1440	673.3	-63.17	-56.56
11	2000	0.2	24	1423	141	44.21	553	1600	1423	694.0	-63.06	-56.83
12	2000	0.2	30	1366	137.1	43.79	556.4	2035	1366	693.5	-62.71	-56.82
13	2000	0.3	18	1879	199.3	40.41	552.4	1333	1879	751.7	-65.48	-57.52
14	2000	0.3	24	1827	237.6	48.41	535	1511	1827	772.6	-65.23	-57.76
15	2000	0.3	30	1762	219.8	30.17	567.4	1929	1762	787.2	-64.92	-57.92
16	2000	0.4	18	2331	253.7	24.22	605.9	942.5	2331	859.6	-67.35	-58.69
17	2000	0.4	24	2134	259.7	57.46	540.1	1370	2134	799.8	-66.58	-58.06
18	2000	0.4	30	2179	274.3	49.93	524.8	1326	2179	799.1	-66.77	-58.05
19	3000	0.2	18	1435	183.7	33.31	617.7	2572	1435	801.4	-63.14	-58.08
20	3000	0.2	24	1380	181.8	68.46	614.8	2306	1380	796.6	-62.80	-58.02
21	3000	0.2	30	1344	181.1	53.13	631.8	2007	1344	812.9	-62.57	-58.20
22	3000	0.3	18	1869	277.2	81.35	571.7	2256	1869	848.9	-65.43	-58.58
23	3000	0.3	24	1762	283.4	68.77	599.2	2433	1762	882.6	-64.92	-58.92
24	3000	0.3	30	1768	263.3	43.55	631	2299	1768	894.3	-64.95	-59.03
25	3000	0.4	18	2303	295.3	62.01	631.4	1734	2303	926.7	-67.25	-59.34
26	3000	0.4	24	2178	322.5	82.66	582.5	2024	2178	905.0	-66.76	-59.13
27	3000	0.4	30	2190	288	80.73	600.8	2080	2190	888.8	-66.81	-58.98

The signal to noise ratio (S/N ratio) of F_{ste} and T_{ste} are calculated according to the following equation:

$$SN = -10\log(\text{MSD}) \tag{5}$$

where MSD is the mean-square deviation for the output characteristic.

The S/N response table and S/N graph for cutting force F_{ste} are given in Table 3 and Figure 6, respectively. The S/N ratio is "the lager the better". It can be seen that the spindle speed has little effect on cutting force. The difference between the maximum S/N ratio and the minimum S/N ratio for spindle speed is only 0.16, and the S/N graph for spindle speed is nearly a horizontal line. Feed rate has the largest influence on cutting force. The difference between the maximum S/N ratio and the minimum S/N ratio for feed rate is 4.05. With increasing feed rate, the S/N ratio decreases apparently, meaning that the corresponding cutting force increases apparently. On the contrary, helix angle has a positive effect on cutting force. With increasing helix angle, the S/N ratio increases as well, meaning that the corresponding cutting force decreases.

Table 3. S/N response table for cutting force.

		Mean S/N ratio (dB)			
Symbol	Parameter	Level 1	Level 2	Level 3	Max-Min
A	N	-65.12	-65.03	-64.96	0.16
B	f_0	-62.90	-65.25	-66.95	4.05
C	H_a	-65.39	-64.93	-64.78	0.61

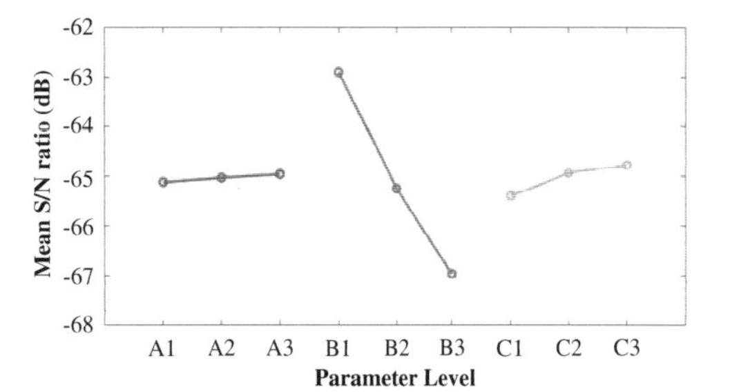

Figure 6. S/N graph for cutting force.

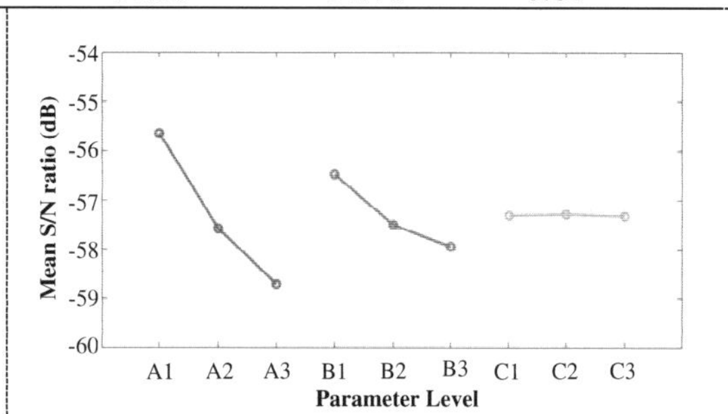

Figure 7. S/N graph for tool temperature.

An analysis of variance (ANOVA) is carried out for cutting force. The results are given in Table 4. The F-value of 623.57 implies the investigated model is significant. There is only a 0.01 percent chance that the F-value of this large could occur due to noise. The p-values for feed rate and helix angle are less than 0.05, indicating that these two parameters have significant effects on cutting force.

Table 4. ANOVA for cutting force.

Symbol	Sum of Squares	DF	Mean Square	F-value	p-value
Model	76.38	6	12.73	623.57	<0.0001
N	0.11	2	0.057	2.79	0.0852
f_0	74.42	2	37.21	1822.61	<0.0001
H_a	1.85	2	0.92	45.29	<0.0001
Error	0.41	20	0.020		
Total	76.79	26			

The S/N response table and S/N graph for tool temperature T_{ste} are given in Table 5 and Figure 7, respectively. It can be seen that both spindle speed and feed rate have negative effects on tool temperature. With larger spindle speed and feed rate, the S/N ratio for tool temperature decreases. The S/N graph for helix angle is nearly a horizontal line, indicating that the helix angle has little effect on the tool temperature. The difference between the maximum S/N ratio and the minimum S/N ratio for helix angle is only 0.04.

Table 5. S/N response table for tool temperature.

Symbol	Parameter	Mean S/N ratio (dB) Level 1	Level 2	Level 3	Max-Min
A	N	-55.65	-57.58	-58.70	3.05
B	f_0	-56.48	-57.50	-57.94	1.46
C	H_a	-57.31	-57.29	-57.32	0.04

The ANOVA for tool temperature is given in Table 6. The F-value of 138.81 implies the investigated model is significant. The p-values for feed rate and helix angle are less than 0.05, indicating that these two parameters are significant. The p-value for helix angle is larger than 0.1, indicating that this parameter is not significant.

Table 6. ANOVA for tool temperature.

Symbol	Sum of Squares	DF	Mean Square	F-value	p-value
Model	53.04	6	8.84	138.81	<0.0001
N	42.92	2	21.46	336.98	<0.0001
f_0	10.11	2	5.06	79.41	<0.0001
H_a	0.006	2	0.003	0.048	0.9530
Error	1.27	20	0.064		
Total	54.31	26			

4. Conclusions

This paper studied the effects of spindle speed, feed rate and helix angle on cutting force and tool temperature in drilling based on a series of finite element simulations. Feed rate was found to be the most influential process parameter on cutting force, while spindle speed was found to be the least influential one. Both spindle speed and feed rate had negative effects on tool temperature, while helix angle had little effect on tool temperature. This paper can provide a basis for suitable selection of process parameters for drilling operations.

References

[1] Lazoglu, I., et al., Thermal analysis in Ti-6Al-4V drilling. CIRP Annals, 2017. 66(1): p. 105-108.
[2] Sun, J. and Y.B. Guo, A comprehensive experimental study on surface integrity by end milling Ti-6Al-4V. Journal of Materials Processing Technology, 2009. 209(8): p. 4036-4042.
[3] Molinari, A., C. Musquar and G. Sutter, Adiabatic shear banding in high speed machining of Ti-6Al-4V: experiments and modeling. International Journal of Plasticity, 2002. 18(4): p. 443-459.
[4] Anand, R.S., et al., Mechanistic modeling of micro-drilling cutting forces. The International Journal of Advanced Manufacturing Technology, 2017. 88(1-4): p. 241-254.
[5] Tuğrul Özel, A.T., Process simulation using finite element method-prediction of cutting forces, tool stresses and temperatures in high-speed flat end milling. International Journal of Machine Tools & Manufacture, 2000. 40(5): 713-738.
[6] Sulaiman, S., A. Roshan and M.K.A. Ariffin. Finite Element Modelling of the effect of tool rake angle on tool temperature and cutting force during high speed machining of AISI 4340 steel. 2013.
[7] Muhammad, R., et al., Finite-Element Analysis of Forces in Drilling of Ti-Alloys at Elevated Temperature. Solid State Phenomena, 2012. 188: p. 250-255.

Modelling the Belt Perforation Process with the Piercing Punch and the Die in the Context of the Construction of the Punching Dies

D Wojtkowiak and K Talaśka

Chair of Basics of Machine Design, Poznan University of Technology, Piotrowo 3 street, 61-138 Poznań, Poland

dominik.wojtkowiak@put.poznan.pl

Abstract. Proper usage of modelling in the design process should result in achieving the effective construction of the machine. However, the methodology of modelling should be adjusted to the specific application. This paper presents the algorithm of modelling the belt perforation process. Authors show the main aspects that should be taken into consideration during the design process of punching dies. Important correlations between construction, process and exploitation parameters are described along with the analysis methods. The main focus of the paper is put on the geometry of the piercing punch, which can significantly change perforation force, tool life and hole quality when it is modified. The paper also analyses the impact of punch velocity and temperature on the force necessary to pierce the material. Additionally, the methodology of determining the mechanical properties of the belt used for further modelling is discussed.

1. Introduction

Modelling of mechanical systems and processes performed by them may provide a lot of helpful data, which can be further used for the design process of modern machines. The most complex set of useful information can be obtained by using the combination of all three known approaches for modelling: deriving analytical models, performing numerical analysis or computer simulations and performing experimental tests. All these methods can be combined in any possible way, e.g. the model may be derived based on the computer simulations, where the computational model is validated by the experimental results or the numerical analysis may be performed based on the analytical model. Choosing the right method depends on many factors, like the object of research interests, its application, the analysed aspect, etc. This indicates that it cannot be generalized and the proper methodology of modelling should be determined for the specified type of the machines. Since nowadays machines tend to be automated, modelling should be performed in the context of both the construction of the machine or its systems and the steering process of the device. In the presented research, authors have focused on modelling within the context of the construction of punching dies for belt perforation, however the methodology derived by the authors will also provide useful information for the design process of its steering system.

In the paper, authors have presented the methodology of modelling the belt perforation process with the piercing punch and the die. As was proved in the following studies [1-3], this method is the most suitable one for multilayer composite belts and has a lot of potential to be modified in the context of lowering perforation force, extending tool life and improving hole quality in perforated belts. All these parameters can be modelled in the function of the piercing punch geometry in order to find the effective construction solution. If the combination of these functions is taken into consideration, the optimization of the tool shape may be performed, as presented in [3]. Since the high divergence of the belt's mechanical properties occur, it is difficult to determine one universal tool for all types of the belt. By using the presented methodology it is possible to design a few sets of tools suitable for various groups of belts. By maintaining the facility of the piercing punch replacement, it is possible to design universal automated machines for belt perforation, which will improve the production process of vacuum conveyor belts.

2. Methodology of modelling belt perforation process

The methodology of modelling the belt perforation process in the context of the construction of the punching dies is presented in figure 1.

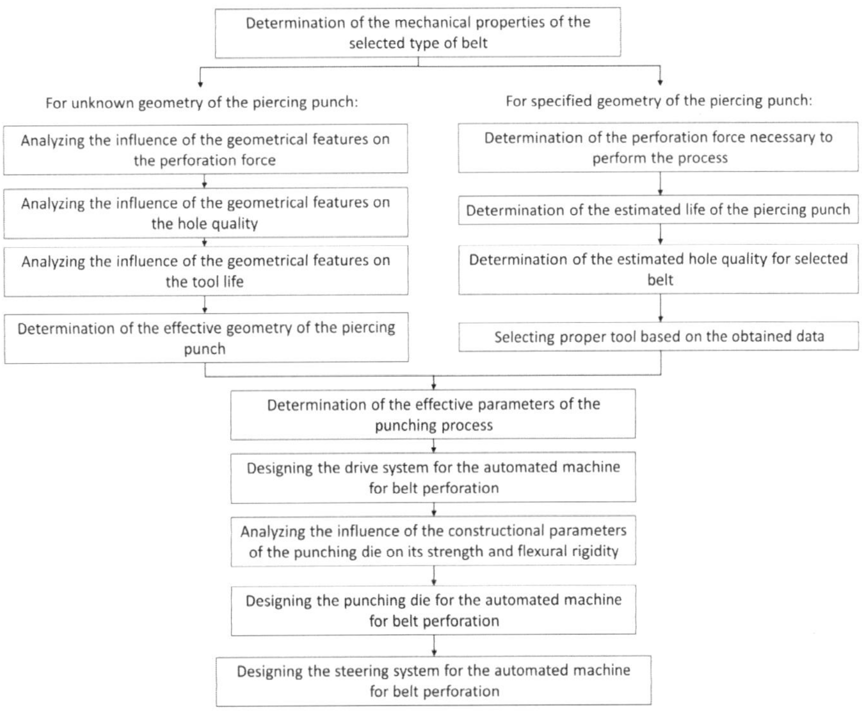

Figure 1. Algorithm of modelling the belt perforation process in the context of the construction of punching dies

It can be observed that in the presented algorithm modelling is used in all stages of designing the machine. At the preliminary stage, mechanical properties of the selected belt type are determined and the decision about designing the tool or selecting the one with specified geometry has to be made. The data obtained in the preliminary stage are used in the elaboration of the detailed technical design. Since the main focus of the research is put on the tool geometry and its influence on perforation force, tool life and hole quality, the majority of modelling occurs at this stage. Depending on the previous decision, modelling can be used either for analysing the above mentioned correlations or applying them in the optimization process to determine the effective shape of the piercing punch. In the next stages, the impact of process parameters like punch velocity or temperature, is analysed and the results are directly used in the design of the drive system. The same method is used for the design of the punching die, in which the results of strength and rigidity analyses are applied in order to achieve the effective construction. Although in the presented algorithm modelling does not cover the stages of manufacturing, prototype-testing and technical documentation preparation, it does indirectly facilitate all of these processes. The generated 3D models of the machine components necessary for FEM analysis can be used for computer aided manufacturing (CAM) and will speed up the preparation of the working and assembly drawings. Thanks to the computer simulations the number of physical prototypes which have to be analysed can be reduced to a minimum. These prototypes will be used only to verify the results obtained in the modelling process.

Based on the above description, it is visible that modelling may improve the design process and the proper methodology for a selected type of the machine can be useful for engineers. In the following sections authors have presented the main assumptions and examples of the analyses for each stage of the modelling.

3. Determination of the mechanical properties of selected types of belts

The first stage of modelling is to establish the entry data which will be used in the design process of the punching die. Since the properties of multilayer polymer composite belts or fibre-reinforced ones can be extremely different [4], the construction features of the designed device may also be characterized with rather high divergence. This indicates that modelling should be performed for selected types of belts or a single group have to be used as the main one. Although the first solution will make it possible to design the universal machine for belt perforation, it will consume a lot of time to perform all the necessary analysis. For that reason it is advantageous to analyse the range of used materials and select the one which will cause the most issues. With this approach, it is possible to spare some time while still maintaining satisfactory effects for other types of belts.

The methodology for determining mechanical properties for selected types of belts is presented in figure 2. Considering the peak perforation force and tool wear, the most rigid belts with increased strength should be considered as the main one. On the other hand, for the quality of the holes elastic light belts or fibre-reinforced ones will cause more problems during the perforation [1]. At this stage the designer has to decide which of the factors is more important and specify the least possible number of belts which will be taken into consideration or analyses which have to be performed in order to achieve the effective punching die construction.

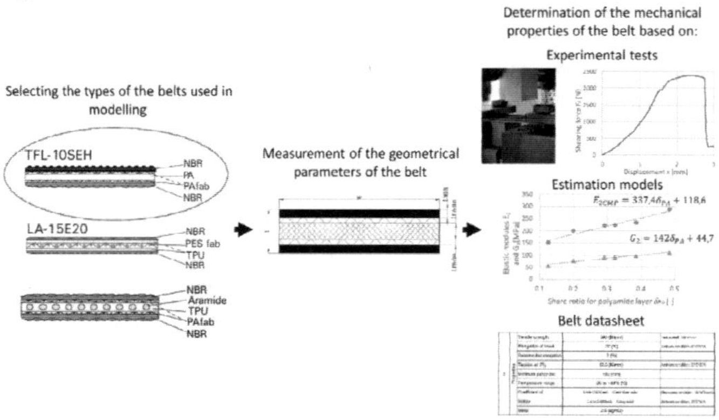

Figure 2. Methodology of determining mechanical properties for selected types of belts [4]

4. Determination of the effective geometry of the piercing punch

The next stage of modelling can be performed in two different ways depending on whether the tool geometry is defined or not. If we want to use the tool which is already designed and manufactured, the main focus will be put on experimental tests. Based on their results, the peak perforation force can be determined for the selected group of belts, which can be later used in the design process of the machine drive system. Additionally, the 3D model of the piercing punch can be made using reverse engineering and then the finite element method (FEM) analysis may be performed to estimate the tool life. The results of the modelling can be further used to create the datasheet for the selected tool. To evaluate its properties and efficiency of the perforation with this tool we can use the shape factor β, which is the ratio of the peak perforation force for the modified profile tool to the force obtained with a flat punch. If a proper database of datasheets for various tools is generated by the manufacturers of perforated belts, the selection of the right tool will be much easier for engineers. The example of the analysis for different punch profiles is presented in figure 3. The obtained results as well as the research presented in [2, 5] proved that the most suitable and the most universal shape of the piercing punch is the one with a spherical bowl.

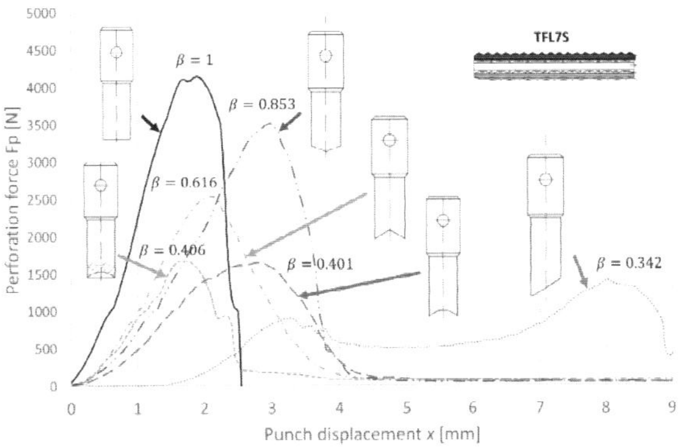

Figure 3. Comparison of experimental characteristics of perforation force in the function of the piercing punch displacement for various punch shapes and the belt TLF7S

In case the geometry of the punch is not specified and should be designed along with the punching die, the methodology of deriving the effective geometrical features of the tools is a combination of all methods of modelling described in the introduction. To derive the correlation between the perforation force and the radius of the spherical bowl *R* or the diameter of the punch *D,* FEM analysis should be used and then the experimental results should verify the computational model. In order to estimate the tool life, the best way is to use the analytical model derived in [3], based on which the correlation between the perforation force to transverse force ratio F_P/F_T and the punch frictional wear can be found. The evaluation of the hole quality can be performed based on the measurement of hole diameters in both perpendicular directions: longitudinal and transverse, which are punched during experimental tests used for perforation force modelling. Based on the obtained results, quality indicators like hole deviation Δ or ovality factor δ can be calculated. Quality can also be graded based on the number of hole defects, i.e. pull-out fibres, uneven edges, etc. All these correlations are used to perform the optimization of the punch geometry [3].

The impact of the selected geometrical features of the piercing punch is presented in figure 4. Based on the results, we can conclude that the differences in perforation force for various bowl radiuses can reach 50%, while the punch-die clearance influence does not exceed 10%. This indicates that when deriving the model for force estimation (figure 5a), the clearance may be neglected and a proper radius of the spherical bowl should be selected (figure 5b). The methodology of developing the FEM model along with the example results is presented in figure 6.

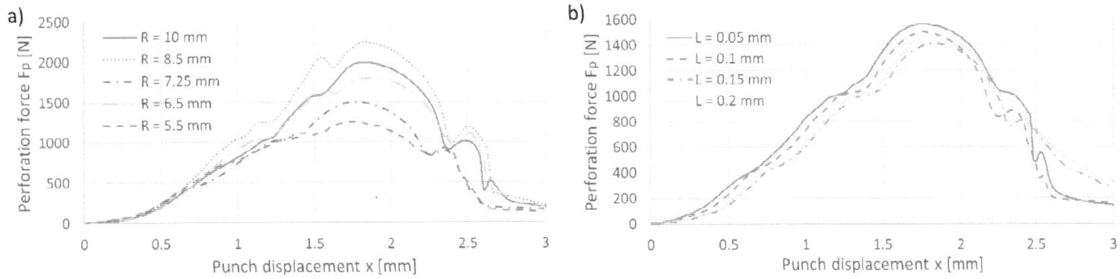

Figure 4. Example of the experimental tests results showing the influence of the tool geometry on the perforation force: a) for various bowl radiuses R, b) for various punch-die clearances

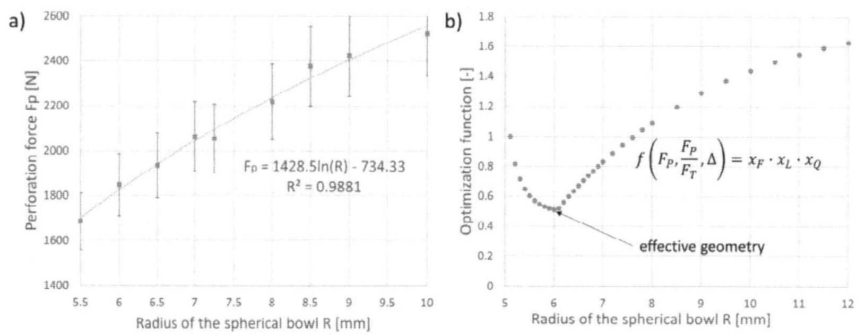

Figure 5. Example of the estimation model for perforation force based on the spherical bowl radius R and the optimization function for TFL10S belt and diameter D = 10 mm

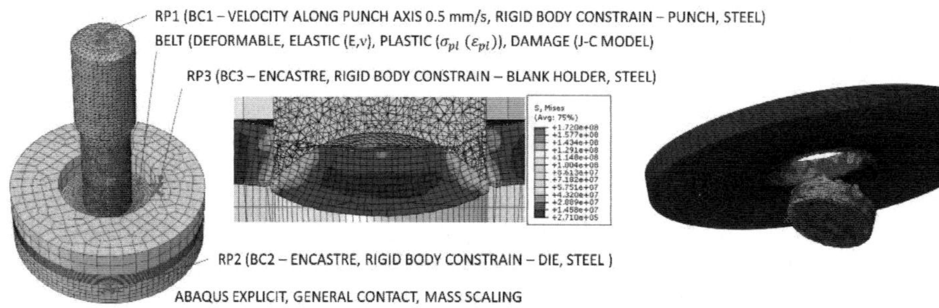

Figure 6. FEM model of belt perforation and the analysis results

5. Determination of the effective parameters of the process

Regardless of whether the punch geometry is specified or has to be designed, the next stage of modelling the belt perforation is to determine the influence of the process parameters on exploitation parameters, i.e. peak perforation force, tool life and hole quality. The most important parameter of the punching process is the velocity of the punch. It has a significant effect on the force necessary to pierce the material, since the dynamics of the process increases with greater velocity. Due to the rheological characteristics of the polymer multilayer composite belts, the dynamic response of the material to the compression generated by the punch varies and affects the peak value of the perforation force. The correlation was proved based on the experimental results and presented in figure 7a.

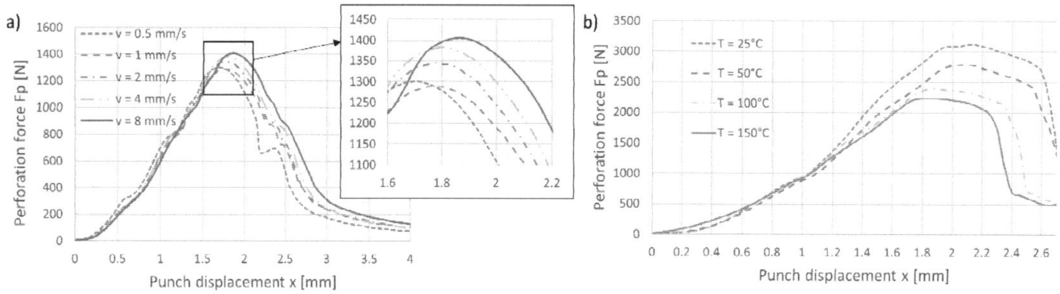

Figure 7. Correlation between the perforation force and process parameters: a) punch velocity, b) punch temperature – experimental results

The tests have shown that although the tendency is kept for different belts, the divergence between the magnitude of the influence may occur. This indicates that a proper analysis of correlation between punch velocity and the above mentioned exploitation parameters should be made. For that purpose, FEM analysis using the J-C material model (as presented in section 5) can also be used. In order to take into consideration the dynamic response of the material, the d_4 in the J-C damage model, which is responsible for the strain velocity impact, should be set as non-zero value. This approach, along with the determination of the constant value, will be presented in the authors' future research papers.

The second parameter which may have some influence on the performance of the punching die and can be steered in the process is the temperature of the piercing punch. In most technological processes the increased temperature is the negative effect of the dissipation of the used energy and may affect the structure or mechanical properties of the machined parts. It is one of the advantages of using the punching for belt perforation instead of drilling or laser cutting [1]. However, as was proved in [6], an increasing temperature of the tool in a limited range may also have a positive effect, especially on hole quality and perforation force. The reason for this phenomenon is softening the material and increased efficiency of hot cutting. This effect can also be analysed using FEM by detailing the J-C model with the constant of the material d_5 or it can be tested during a series of experiments as presented in figure 7b.

6. Analysing the construction parameters of the punching die

The last stage in which modelling is used during the punching die design is analysing the construction parameters of the punching die. In this group we can distinguish a kinematic and dynamic analysis connected with the drive system of the machine, strength analysis of the main components of the device and rigidity analysis connected with maintaining a proper alignment between the punch and the die. Most of them are performed as FEM analysis, in which the previously performed perforation force plays a role of the loading. Based on the obtained results, a proper drive can be selected or dimensions of the parts can be calculated. Additionally, mass reduction is possible using the topology optimization of the construction (mainly of the head-punch and the base-die plates). Since it is very hard to verify the obtained results, it is advantageous to perform the analysis of the meshing influence, which should eliminate coarse errors.

7. Summary

The methodology presented in the paper can be used to design more effective constructions of punching dies. One of the advantages of the presented algorithm is combining all modelling methods (analytical, simulational and experimental one), which makes it possible to perform an efficient process of design. As can be observed, a seemingly very simple process of belt punching causes a lot of trouble for perforated belts manufacturers. With a detailed analysis it is possible to discover the complex nature of the process and achieve the construction which is even 60-70% more efficient. All these aspects prove that the presented research provides a lot of useful data, which can be applied by the engineers connected with the selected branch of industry and should facilitate their work.

References

[1] Wojtkowiak D, Talaśka K, Malujda I and Domek G 2017 *Mechanik* **12** 1138-1142
[2] Wojtkowiak D, Talaśka K, Malujda I and Domek G 2018 *Int. J. Adv. Manuf. Technol.* **98,** 9-12, 2539-2561
[3] Wojtkowiak D and Talaśka K 2019 (in press) *Int. J. Adv. Manuf. Technol.* doi: 10.1007/s00170-019-03746-7
[4] Talaśka K and Wojtkowiak D 2018 *MATEC Web of Conferences* **157** 02052
[5] Wojtkowiak D and Talaśka K 2019 *MATEC Web of Conferences* **254** 02001
[6] Wojtkowiak D, Talaśka K, Malujda I and Domek G 2018 *MATEC Web of Conferences* **157** 01022

Machining Grooves in Silica Glass by Using Picosecond Lasers

J K Jiao[1], J A Zhang[2], W W Zhang[1], R Popvawe[3]

[1] Ningbo Institute of Materials Technology and Engineering, CAS, Ningbo 315201 China
[2] Ningbo CrysDiam Industrial Technology Co., Ltd. Ningbo 315201 China
[3] Fraunhofer Institute for Laser Technology, Aachen, Germany

jiaojunke@nimte.ac.cn

Abstract. A picosecond laser with 1064nm wavelength was used to machine grooves in fused silica in this paper. A surface texturing method was applied to increase the laser absorption. The relationship between the width, the depth of the groove and the laser parameters such as laser power, scanning speed, number of scanning cycles was studied. The machining quality of the groove was detected by the laser scanning confocal microscopy. It showed that there were no micro-cracks on the grooving surface, and the surface roughness Ra<1.6um. The width, the depth and the shape of the groove could be controlled accurately by changing the laser parameters. And, a "V-shape" groove with 1mm width, 1mm depth was obtained with the as the frequency, the laser power, the scanning velocity and the number scanning cycle was fixed at 400KHz, 141W, 5000mm/s and 12, respectivity.

1. Introduction

Fused silica is an important material widely used in aerospace, nuclear, electrics, optics, mechanical engineering. However, it is difficult to micro-machine with traditional methods because of its hardness and brittle characters. With the development of laser technology, laser has been applied in machining brittle materials. Jiao[1,2] cut glass plates using CO_2 lasers with fracture control method, and the glass can be cut with a high quality using dual laser beam method. The temperature and the thermal stress was simulated in the cutting process by Tian[3] and Wei[4]. The result showed that the cutting quality was affected by the temperature distribution greatly. Brugan [5] used simultaneous CO_2 lasers to cut alumina ceramics and Wang [6,7] cleaved glass sheets by using a line-shape laser beam with controlled fracture and volumetric heat absorption.

As mentioned above, CO_2 laser was used to machine glass in most of these studies because of its high absorption coefficient for glass. As we all know, the fused silica was transparent to light with wavelength around 1000nm. Long pulse laser or CW laser around this wavelength has a low efficiency in processing glass material. However, the short or ultra-short pulse laser has a great talent in fused silica machining [8]. In this study, a surface texturing method was applied to increase the laser absorption coefficient. Picosecond laser with 1064nm wavelength was used to machining grooves in the fused silica. The relationship between the width, the depth of the groove and the laser parameters was researched.

2 Experimental design

In this study, a picosecond laser by EdgeWave was used. The maximum laser power is 300W at 2-20MHz, the wavelength is 1064nm, the frequency can be adjusted from 400KHz to 20MHz. The laser beam is reflected into a laser scanner, the focused laser diameter is ~75 microns. The micro-processing path was controlled by the laser scanner, and the maximum speed is10m/s. The experimental setup is shown in Figure 1 and Figure 2(a).

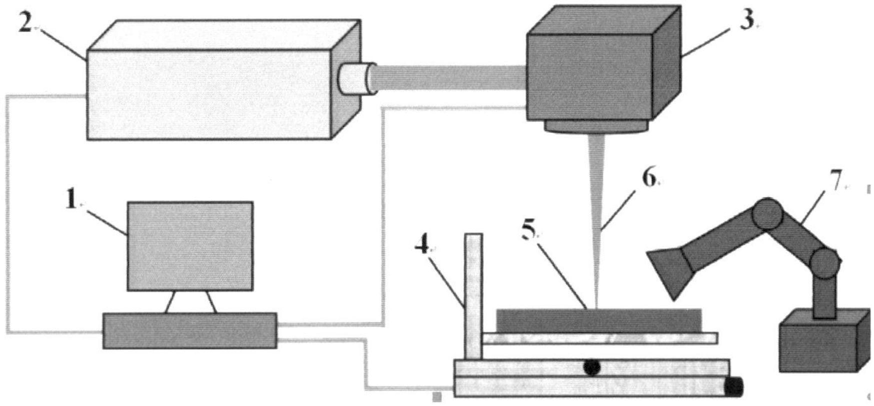

Figure.1 Experimental setup (1.Control system, 2.Picosecond laser, 3.laser scanner, 4.XYZ motion system, 5.Substrate, 6. Laser beam, 7.Dust exhaustion apparatus)

To measure the cross section and the roughness of these grooves, a Keyence confocal laser scanning microscope (Figure 2(b)) was used to examine the characteristics of the microgrooves, and the width, depth, cross section, roughness and the taper of the grooves can be measured conveniently. This microscope has 110 nanometer XY-plane resolution and 10 nanometer Z-resolution, a 50 mm by 50 mm zone can be scanned automatically to get the 3D geometry of the machined features. The optical magnification is 3000X, with digital magnification, the final magnification reaches 24000X.

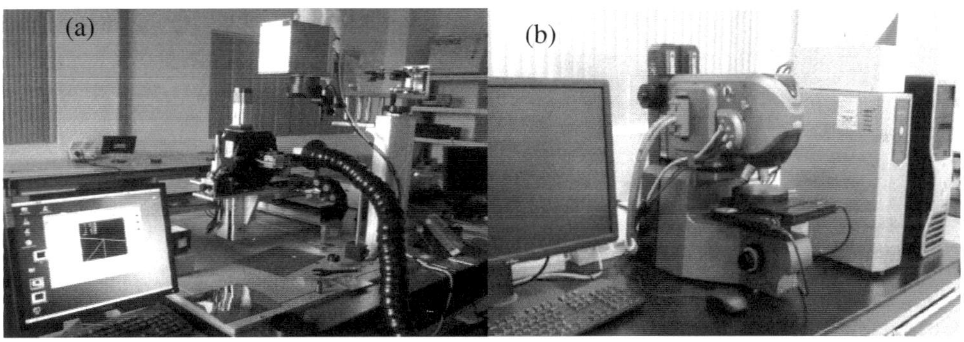

Figure.2 (a) The experimental setup and (b) the confocal scanning microscope

In the experiment, the laser beam scans N lines on the substrate surface for one scanning layer (Figure 3). The pulse overlap of two scanning is about 1/4 of laser spot size (Δd=20um), which is the best value obtained through previous experiments. For one scanning cycle, there are N scanning layers. The laser beam scans N times for the first layer, and N-1 times for the second layer, then N-2 times for the third layer, and so on, 1 time for the last layer. The schematic diagram of this scanning method was shown in Figure 3.

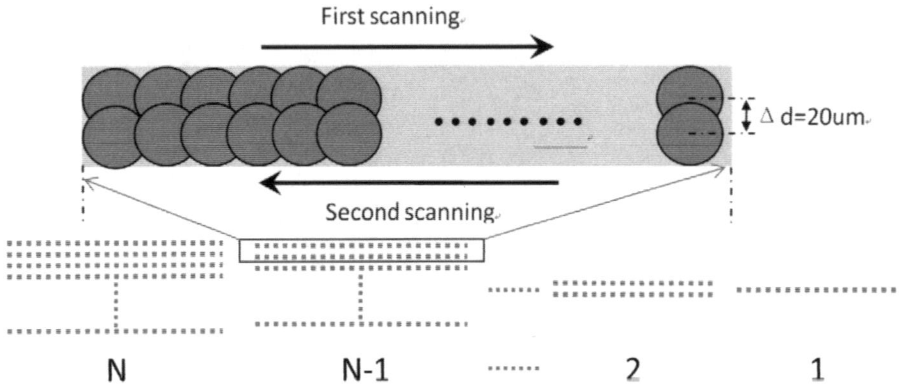

Figure.3 Scanning method for V-shape groove in one scanning cycle

3 Result and discussions

For lasers with 1064nm wavelength, more than 90% laser energy transmits out from fused silica [9]. In order to increase the absorption coefficient, the fused silica surface was polished with the abrasive paper to increase the surface roughness in this study. The roughness of the polished surface was measured by using a confocal laser scanning microscope. The measured result showed that there are lots of micro-pits on the surface and the roughness is about 9.2um (Figure 4). The absorption coefficient can be increased because of these micro-pits and roughness surface.

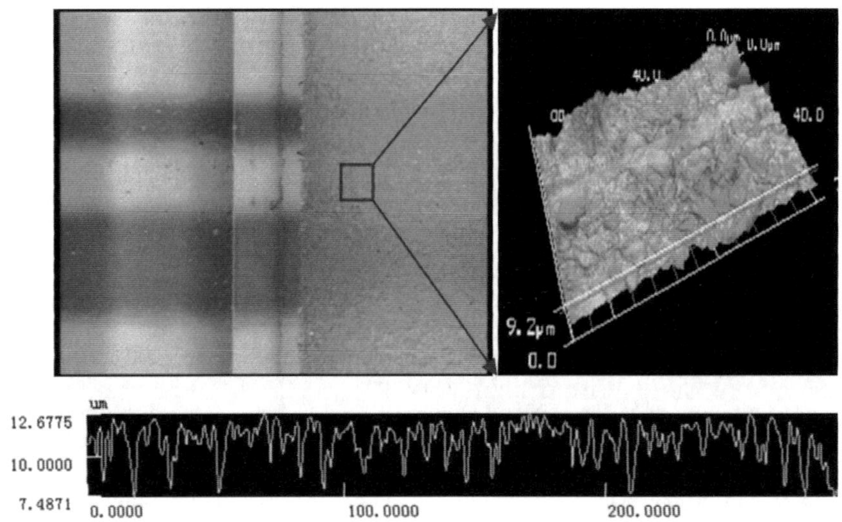

Figure.4 Roughness of fused silica polished by abrasive paper

After polished with the abrasive paper, the fused silica surface absorption coefficient to laser increases greatly. The material was vaporized under the high power picosecond laser, and micro-grooves were achieved. However, the fused silica without any treatment, the absorption coefficient is very low in the surface, and some cracks were produced in the substrate, especially on the laser-in surface. One traditional method to improve the absorption coefficient is coating high absorption material on the fused silica surface. In this study, a material with high absorption coefficient to 1064nm laser was coated on the surface. With this method, more than 85% laser energy was absorbed. With this method, lots of micro cracks were produced on the surface because of thermal stress accumulation. With these micro cracks, the surface roughness was improved and more laser energy could absorbed by fused silica. The glass material was vaporized by the high power laser pulse in this case, and a new surface with micro-pits was produced, the laser absorption coefficient increases again. With this method, the fused silica material was vaporized layer by layer, and micro groove was produced. However, lots of cracks were produced on the edge of the groove (as shown in Figure 5) because of a great of thermal energy accumulated on the surface in a short of time as the laser beam irradiating on it. On the other hand, the fused silica polished with abrasive paper could be machined by picosecond laser without any micro cracks. As shown in Figure 6, a groove with 900um width and 620um depth was produced, and the roughness on the groove bottom Ra=0.82um, which is smaller than that of polished with abrasive paper, and it satisfy most of applications in industry.

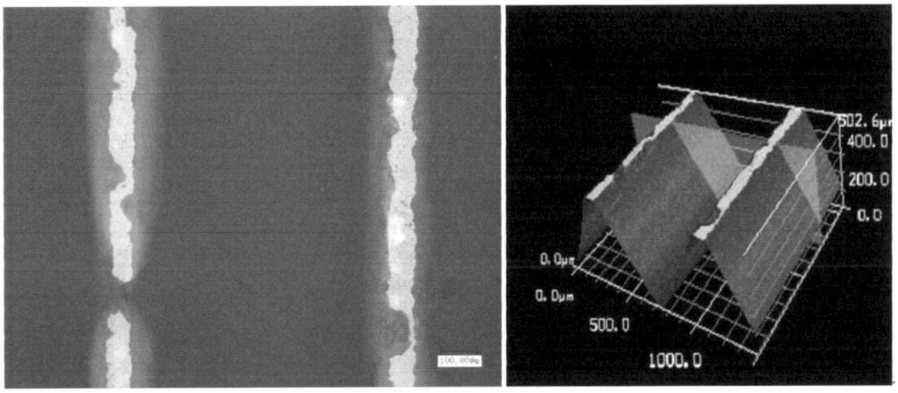

Figure.5 Micro-grooves in fused silica with coating high absorption material

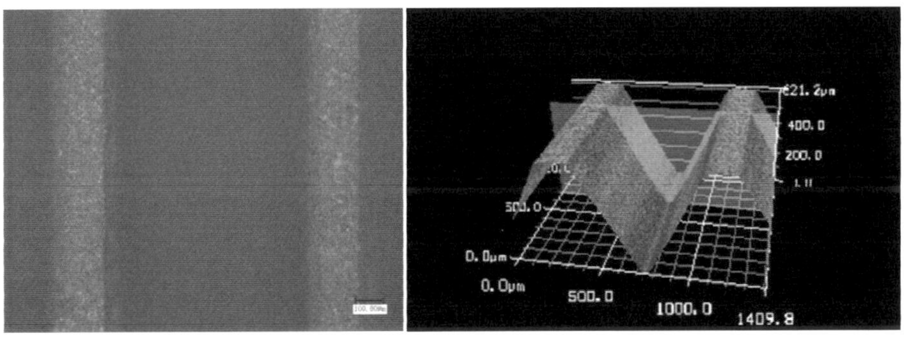

Figure.6 Micro-grooves in fused silica with abrasive paper polishing

In our previous research, it was found that the material removal efficient increases firstly and then decreases with the increasing of the frequency, and the most efficient material removal frequency was fixed at 400KHz. Under this frequency, the maximum laser power is 141W. The scanning velocity is chosen as 5000mm/s, which is the best value under 141W laser power. With these laser parameters, the shape of the micro-groove under different number of scanning cycle was researched. In this study, the number scanning cycle was chosen as 7, 8, 9, 10, 12, separately. As shown in Figure 7, d indicates the 1/2 width of the groove, h is the depth of the groove, and s is the 1/2 cross section area of the groove.

As shown in Fig.7, with the increasing of the scanning cycle number, the depth of the groove increases firstly, and then nearly keeps in a constant value after 9 scanning cycles, which is about 570um. It because the vaporized material could not escape from the groove as the groove depth reached a specific value. The width of the groove increases with the increasing of scanning cycles, which is about 1020um for 12 scanning cycles. Fig.8 gives the changing of groove taper with scanning cycle number. The angle

IOP Conf. Series: Materials Science and Engineering **647** (2019) 012014 doi:10.1088/1757-899X/647/1/012014

is smaller, the taper of the groove is larger. As shown in the Figure 8, the taper of the groove becomes larger with increasing of the scanning cycle number.

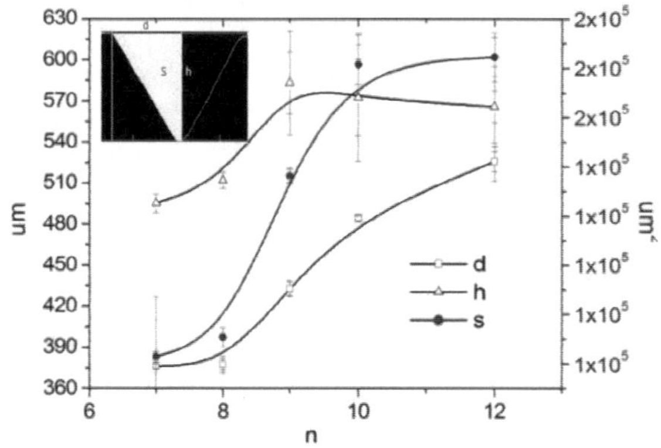

Figure.7 Shape of the micro-groove

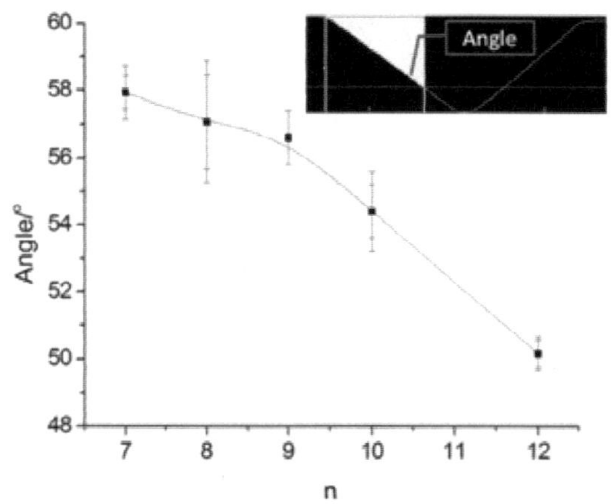

Figure.8 The groove taper changing with scanning cycle number

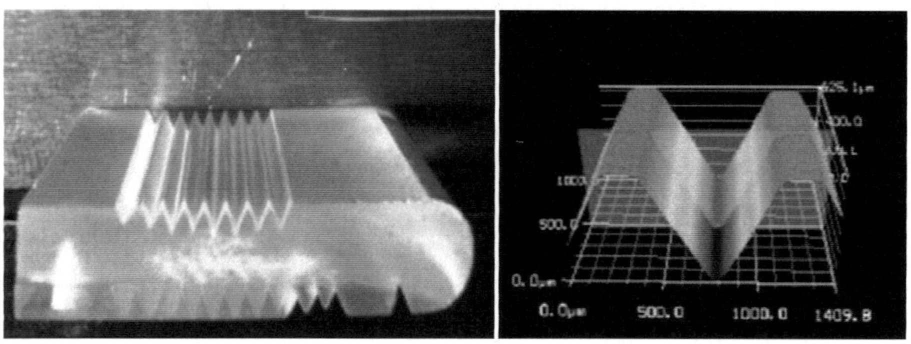

Figure.9 V-shape grooves produced in the fused silica

Grooves with different shape could be obtained by controlling the scanning cycle number, scanning line number for different layer and the laser parameters. As shown in Figure 9, V-shape (h=0.6mm, d=0.5mm) groove was achieved under the scanning parameters as shown in table1.

Table 1 Laser parameters for groove machining

Laser power	Frequency	Scanning velocity	number of scanning cycle
141W	400KHz	5m/s	12

4 Conclusion

The absorption coefficient of fused silica for 1064nm laser could be improved by texturing method. After polishing with abrasive paper, the fused silica could be machined by picosecond laser without any micro cracks. And the roughness of the groove bottom was Ra=0.82um.The shape of the groove could be controlled accurately by controlling the laser parameters and the cycle number of scanning.

Acknowledgement

The work was funded by the Ningbo international cooperation project (2017D10007), the NSFC-Zhejiang Joint Fund for the Integration of Industrialization and Information (U1609208), and Youth Innovation Promotion Association (2017343), and National Commercial Aircraft Manufacturing Engineering Technology Research Center Innovation Fund (COMAC-SFGS-2017-36734).

References

[1] J K Jiao and X B Wang 2009 *Optics and Laser in Engineering* **47**(7-8):860-864
[2] J K Jiao and XB Wang 2008 *Optics and Laser Technolog* **40**(2): 297-301
[3] C Y Wei H B He and Z Deng 2005 *Optical Engineering* **44**(4):1~4
[4] W X Tian Wilson K. and S. Chui 2004 *Optics and Laser Technology* **36**(2):131~137
[5] P Brugan G Cai and R Akarapu 2006 *Journal of Laser Applications* **18**(3):236~241
[6] J F Tu and P E B Riley 2013 *Optics and Lasers in Engineering* **51**(6):696-706
[7] Y Z Wang and J Lin 2007 *Optics and Laser Technolog* **39**(9):892~899
[8] L J Yang and Y Wang 2010 *In. J. Machine Tools and Manuf.* **50**(10): 849–859
[9] E Vanagas and J Kawai 2004 *Society of Photo-Optical Instrumentation Engineers* **3**(2):358~363

IOP Conf. Series: Materials Science and Engineering **647** (2019) 012015 doi:10.1088/1757-899X/647/1/012015

Grinding Deviation Analysis of Offset Face Gear Based on Involute Disc Wheel

Yang Zhong*, Jie Wang, Tingyu Zhang, Tianhe Pei, Yuan Gao and Wenhao Yao

Aerospace research institute of materials & processing technology,Beijing,China

* E-mail: zhongyang010@163.com

Abstract. Offset face gear drive has shown many unique advantages in the area of crossed-axes transmission. In this paper, deviation analysis of an offset face gear using a involute disc wheel as a cutting tool has been discussed to achieve precised machining.Firstly, equations of involute disc wheel and offset face gear are derived based on meshing theory.Then, a grinding machine tool is designed and equations of offset face gear containing machining deviation are derived.At last,figures of deviation cloud of face gear surface are created with MATLAB and importance of different machining deviation is discussed.The results show that the deviation Δz along the axis of grinding wheel is the most influential deviation,while Δy along the axis of face gear is the least one.

1.Introduction

Face gear drive is a face gear drived with a spur pinion wherein their axes are intersected, which has unique advantages such as light weight, compact structure and strong bearing capacity. NASA, Boeing, and North Star have verified the unique advantages of face gear in engineering research. [1-3]. Scholars have done lots of research on precise manufacturing, profile modification and meshing analysis of orthogonal face gears[4-8]. However, little research has been done on offset face gears.

Offset face gear drive is a face gear drive where axes of face gear and spur pinion are not crossed. In this paper, grinding process of offset face gear using involute disc grinding wheel has been studied. Tooth surface equations containing machining deviation are derived and figures of deviation cloud are created with MATLAB, which provide a theoretical support for precise machining of offset face gear.

2.Tooth Surface Equation of Offset Face Gear

The tooth surface of offset face gear is formed by a complex generating motion(figure 1) of involute disc grinding wheel. Fixed coordinate systems $S_{s0}(x_{s0},y_{s0},z_{s0})$ and $S_{20}(x_{s20},y_{20},z_{20})$ represent the initial position of grinding wheel and face gear respectively,where E is the offset distance and $x_{s20}=x_{s0}+E$.Moving coordinate system $S_s(x_s,y_s,z_s)$.$S_2(x_{s2},y_{s2},z_{s2})$are fixed to grinding wheel and face gear respectively. When the grinding wheel rotates around the axis z_s to the angle of φ_s at the angular speed of ω_s, the face gear rotates around the axis z_2 to the angle of φ_2 at the angular speed of ω_2 accordingly. The tooth profile of offset face gear $\mathbf{r}_2$ can be expressed as equation 1.

$$\mathbf{r}_2(u_s,\theta_{ks},\varphi_s)=\mathbf{M}_{2s}(\varphi_s)\mathbf{r}_s(u_s,\theta_{ks}) \tag{1}$$

Coordinate transformation matrix $\mathbf{M}_{2s}$ can be expressed as equation 2.

$$\mathbf{M}_{2s} = M_{220} M_{20s0} M_{s0s} = \begin{bmatrix} \cos\varphi_2\cos\varphi_s & -\cos\varphi_2\sin\varphi_s & -\sin\varphi_2 & -E\cos\varphi_2 \\ -\sin\varphi_2\cos\varphi_s & \sin\varphi_2\sin\varphi_s & -\cos\varphi_2 & E\sin\varphi_2 \\ \sin\varphi_s & \cos\varphi_s & 0 & 0 \\ 0 & 0 & 0 & 1 \end{bmatrix} \tag{2}$$

The involute profile of grinding wheel can be expressed as equation 3.

$$\mathbf{r}_s(u_s,\theta_{ks}) = \begin{bmatrix} \pm r_{bs}[\sin(\theta_{os}+\theta_{ks}) - \theta_{ks}\cos(\theta_{os}+\theta_{ks})] \\ -r_{bs}[\cos(\theta_{os}+\theta_{ks}) + \theta_{ks}\sin(\theta_{os}+\theta_{ks})] \\ u_s \\ 1 \end{bmatrix} \quad k = (\gamma,\beta) \tag{3}$$

where r_{bs} is radius of base circle, θ_{os} is initial angle, us and θ_{ks} are variables of involute, and "±"stand for γ,β respectively, as shown in figure 2.

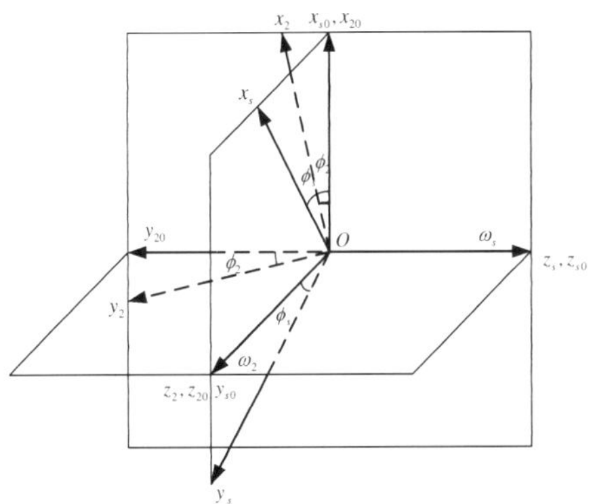

Figure 1. Coordinate systems of offset face gear generating process

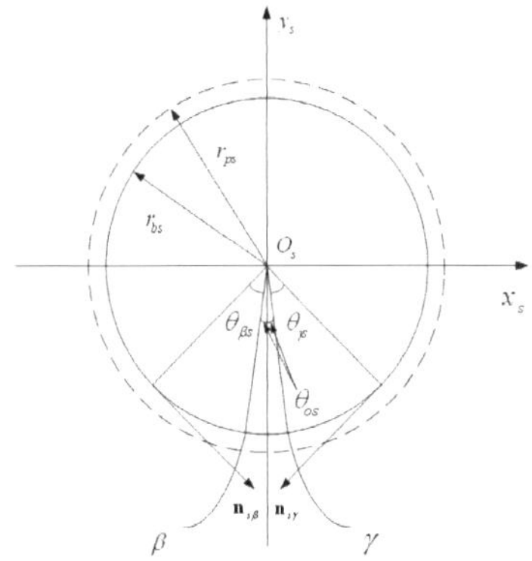

Figure 2. Involute profile of grinding wheel

According to the meshing theory, the unit normal vector of involute tooth profile n_s and the velocity of meshing point $v_s^{(s,2)}$ in ordinary system Ss are calculated. Substituting $n_s \cdot v_s^{(s,2)}=0$ into Eq. 3 to eliminate u_s ,the tooth profile of offset face gear can be given as equation 4.

$$\mathbf{r}_2(\varphi_s,\theta_{ks}) = \begin{bmatrix} r_{bs}\left[\cos\varphi_2(\sin\xi_{ks}\ \mathrm{m}\theta_{ks}\cos\xi_{ks}) - \dfrac{\sin\varphi_2}{m_{2s}\cos\xi_{ks}}\right] - E\cos\varphi_2 \\ -r_{bs}\left[\sin\varphi_2(\sin\xi_{ks}\ \mathrm{m}\theta_{ks}\cos\xi_{ks}) + \dfrac{\cos\varphi_2}{m_{2s}\cos\xi_{ks}}\right] + E\sin\varphi_2 \\ -r_{bs}(\cos\xi_{ks} \pm \theta_{ks}\sin\xi_{ks}) \\ 1 \end{bmatrix} \tag{4}$$

where $\xi_{ks} = \varphi_s \pm (\theta_{ks} + \theta_{os})$,and "±"stand for γ,β respectively

3.Machining deviation analysis of offset face gear

A face gear grinding machine using involute disc wheel has been designed based on generating theory, as shown in figure 3.

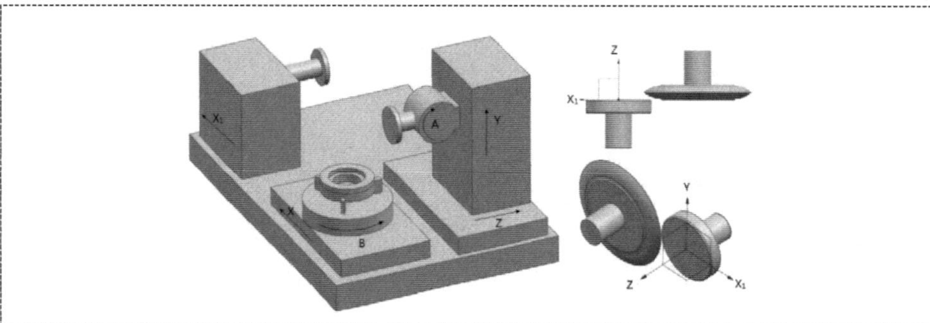

Figure 3. Structure of a face gear grinding machine

The grinding wheel can move along Y and Z axes and rotate around A axis, meanwhile, the workpiece can move along X axis and rotate around B axis. Generation of face gear is achieved by 5-axes(X/Y/Z/A/B) linkage. The dresser is a diamond roller which moves along X_1 axis, and dressing process is achieved by 3-axes(X_1/Y/Z) linkage. When dressing, the trajectory of the roller is the offset curve of the involute.

The deviation of offset distance E and involute profile of grinding wheel can affect the accuracy of tooth profile of offset face gear.In this paper, equations of offset face gear containing machining deviation are derived.Figures of point cloud of offset face gear containing deviation are created by MATLAB and relationship between machining deviation and tooth surface deviation is analyzed,which provides theoretical basis for error compensation.

Parameters of the offset face gear for numerical example are as shown in table 1.

Table 1. Design parameters of face gear

Module	m=5mm
Number of teeth of the shaper	N_s=20
Number of teeth of the face-gear	N_2=41
Pressure angle of driving side and coast side	α=20°
Shaft angle	γ_m=90°
Offset distance	E=25mm
Inner radius of the face-gear	r_2=112mm
Outer radius of the face-gear	r_1=130mm

(1) deviation of offset distance ΔE

Equation of face gear tooth surface containing deviation of offset distance ΔE is given as equation 5.

$$\mathbf{r}_2(\varphi_s,\theta_{ks})=\begin{bmatrix} r_{bs}\left[\cos\varphi_2(\sin\xi_{ks}\mp\theta_{ks}\cos\xi_{ks})-\dfrac{\sin\varphi_2}{m_{2s}\cos\xi_{ks}}\right]-(E+\Delta E)\cos\varphi_2 \\ -r_{bs}\left[\sin\varphi_2(\sin\xi_{ks}\mp\theta_{ks}\cos\xi_{ks})+\dfrac{\cos\varphi_2}{m_{2s}\cos\xi_{ks}}\right]+(E+\Delta E)\sin\varphi_2 \\ -r_{bs}(\cos\xi_{ks}\pm\theta_{ks}\sin\xi_{ks}) \\ 1 \end{bmatrix} \quad (5)$$

where $\xi_{ks}=\varphi_s\pm(\theta_{ks}+\theta_{os})$

The deviation of face gear surface is uniformly distributed along the direction of tooth height(along z axis in figure 4, the same below), and smaller near outer circumference and larger near inner circumference along the direction of tooth width (along y axis in figure 4, the same below). The signal of deviation of right tooth(β)is the same as ΔE ,but opposite to deviation of left tooth(γ).When ΔE=±0.1mm, maximum of deviation of face gear surface δ_E is about ±0.085mm and minimum is about ±0.07mm, as shown in figure 4.

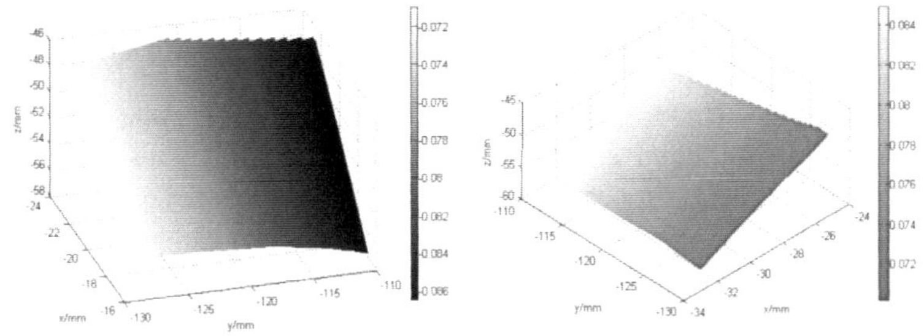

ΔE=0.1mm(surface γ) ΔE=0.1mm(surface β)

Figure 4. Deviation of face gear surface δ_E containing deviation ΔE

(2) deviation of dressing along x_l axis

When there is a deviation Δx along x_l axis during dressing process, the involute profile containing deviation Δx can be given as equation 6.

$$\mathbf{r}_s(u_s,\theta_{ks})=\begin{bmatrix} \pm r_{bs}[\sin(\theta_{os}+\theta_{ks})-\theta_{ks}\cos(\theta_{os}+\theta_{ks})]\pm\Delta x\cdot\sin\dfrac{\pi}{N_s} \\ -r_{bs}[\cos(\theta_{os}+\theta_{ks})+\theta_{ks}\sin(\theta_{os}+\theta_{ks})]+\Delta x\cdot\cos\dfrac{\pi}{N_s} \\ u_s \\ 1 \end{bmatrix} \quad k=(\gamma,\beta) \tag{6}$$

Equation of face gear tooth surface containing deviation Δx can be given as equation 7

$$\mathbf{r}_2(\varphi_s,\theta_{ks})=\begin{bmatrix} r_{bs}\left[\cos\varphi_2(\sin\xi_{ks}\mp\theta_{ks}\cos\xi_{ks})\right]\pm\Delta x\sin(\dfrac{\pi}{N_s}\mp\varphi_s)\cos\varphi_2\text{-}E\cos\varphi_2\text{-}\delta_s\sin\varphi_2 \\ -r_{bs}\left[\sin\varphi_2(\sin\xi_{ks}\mp\theta_{ks}\cos\xi_{ks})\right]+\Delta x\sin(\varphi_s\mp\dfrac{\pi}{N_s})\sin\varphi_2+E\sin\varphi_2\text{-}\delta_s\cos\varphi_2 \\ -r_{bs}(\cos\xi_{ks}\pm\theta_{ks}\sin\xi_{ks})+\Delta x\cos(\dfrac{\pi}{N_s}\mp\varphi_s) \\ 1 \end{bmatrix} \tag{7}$$

where $\xi_{ks}=\varphi_s\pm(\theta_{ks}+\theta_{os})$, $\delta_s=\dfrac{r_{bs}-\Delta x\cdot\cos[\dfrac{\pi}{N_s}+(\theta_{os}+\theta_{ks})]}{m_{2s}\cdot\cos\xi_{ks}}$

The deviation of face gear surface is uniformly distributed along the direction of tooth width, and smaller near tooth root and larger near tooth top along the direction of tooth height. The signal of deviation of tooth surface is the same as Δx. When Δx=±0.1mm, maximum of deviation of face gear surface δ_x is about ±0.075mm and minimum is about ±0.05mm,as shown in figure 5.

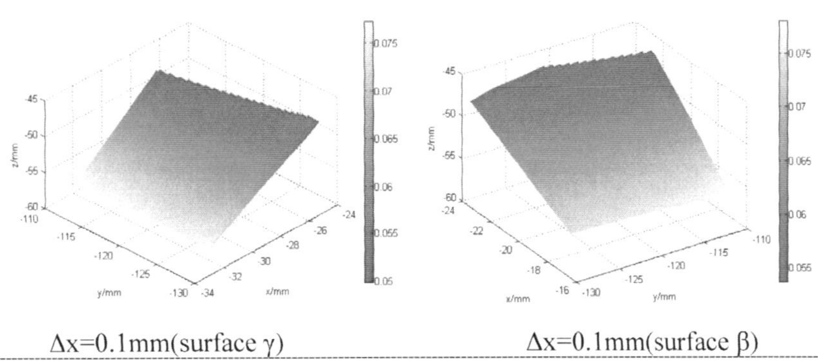

Δx=0.1mm(surface γ) Δx=0.1mm(surface β)

Figure 5. Deviation of face gear surfaceδ_x containing deviationΔx

(2) deviation of dressing along y axis

When there is a deviation Δy along y axis during dressing process,the deviation issue can be simplified to a plane geometry issue as shown in figure 6,where O represents the theoretical center of grinding wheel, O_1 represents the theoretical center of dresser, P represents the theoretical contact point between dresser and grinding wheel, O_1' represents the center of dresser considering deviation Δy, and P' represents the contact point between dresser and grinding wheel considering deviation Δy. The deviation of involute profile of grinding wheel δ_y, which is caused by dressing deviation Δy, can be approximatively given as equation 8.

$$\delta_y = |OP'| - |OP| = |OO_1'| - |OO_1| = \frac{(\Delta y)^2}{|OO_1'|+|OO_1|}. \tag{8}$$

In general, $\Delta y << |OO_1'| + |OO_1|$, so$\delta_y$ can be considered as a a higher order infinitesimal of Δy, which is a insensitive deviation.

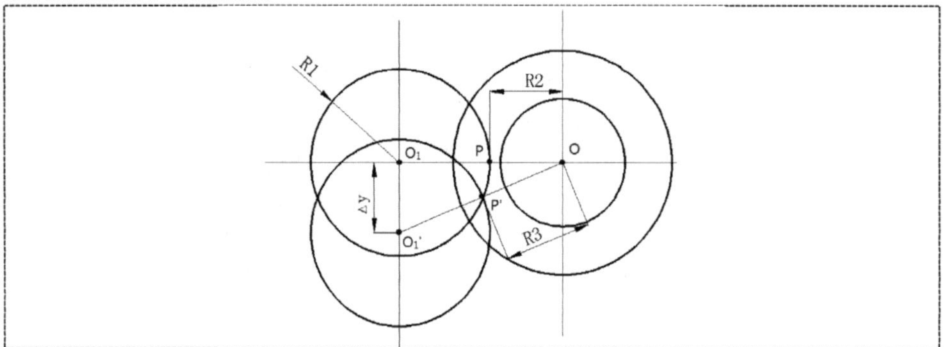

Figure 6. Analysis of deviaiton Δy

(4) deviation of dressing along z axis

When there is a deviation Δz along z axis during dressing process, the involute profile containing deviation Δz can be given as equation 9.

$$\mathbf{r}_s(u_s,\theta_{ks}) = \begin{bmatrix} \pm r_{bs}[\sin(\theta_{os}+\theta_{ks}) - \theta_{ks}\cos(\theta_{os}+\theta_{ks})] \pm \Delta z g\cos\frac{\pi}{N_s} \\ -r_{bs}[\cos(\theta_{os}+\theta_{ks}) + \theta_{ks}\sin(\theta_{os}+\theta_{ks})] + \Delta z g\sin\frac{\pi}{N_s} \\ u_s \\ 1 \end{bmatrix} \qquad k=(\gamma,\beta) \tag{9}$$

The equation of face gear tooth profile containing deviation Δx can be given as equation 10.

$$\mathbf{r}_2(\varphi_s,\theta_{ks}) = \begin{bmatrix} r_{bs}\left[\cos\varphi_2(\sin\xi_{ks}\ \mathrm{m}\theta_{ks}\cos\xi_{ks})\right] \pm \Delta z\cos(\frac{\pi}{N_s}\pm\varphi_s)\cos\varphi_2 - E\cos\varphi_2 - \zeta_s\sin\varphi_2 \\ -r_{bs}\left[\sin\varphi_2(\sin\xi_{ks}\ \mathrm{m}\theta_{ks}\cos\xi_{ks})\right]\mathrm{m}\Delta z\cos(\frac{\pi}{N_s}\pm\varphi_s)\sin\varphi_2 + E\sin\varphi_2 - \zeta_s\cos\varphi_2 \\ -r_{bs}(\cos\xi_{ks}\pm\theta_{ks}\sin\xi_{ks}) + \Delta z\sin(\frac{\pi}{N_s}\pm\varphi_s) \\ 1 \end{bmatrix} \tag{10}$$

where $\xi_{ks} = \varphi_s \pm (\theta_{ks}+\theta_{os})$ and $\zeta_s = \dfrac{r_{bs} - \Delta z g\sin[\frac{\pi}{N_s} - (\theta_{os}+\theta_{ks})]}{m_{2s} g\cos\xi_{ks}}$

The deviation of face gear surface is uniformly distributed along the direction of tooth width, and smaller near tooth root and larger near tooth top along the direction of tooth height. The signal of deviation of tooth surface is the same as Δz. When $\Delta z=\pm 0.1$mm, maximum of deviation of face gear surface δ_z is about ± 0.10mm and minimum is about ± 0.08mm, as shown in figure 7.

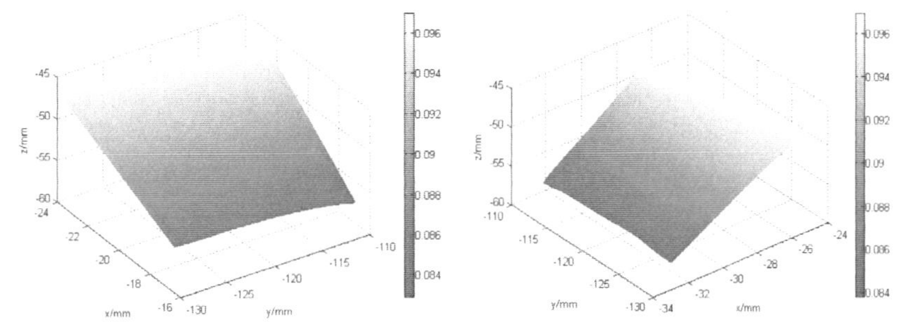

Δz=0.1mm(surface γ) Δz=0.1mm(surface β)

Figure 7. Deviation of face gear surfaceδ_z containing deviation Δz

By adjusting the value of deviation parameters $\Delta E.\Delta x.\Delta z$ from [-0.1,0.1], a further simulation has been carried out (figure 8.) which draws conclusions as follows:

(1) There is approximately an linear relationship between deviation of face gear surface δ and deviation of grinding process Δ.

(2) The importance of deviation of grinding process in descending order is Δz, ΔE, Δx and Δy, and Δy is a insensitive deviation.

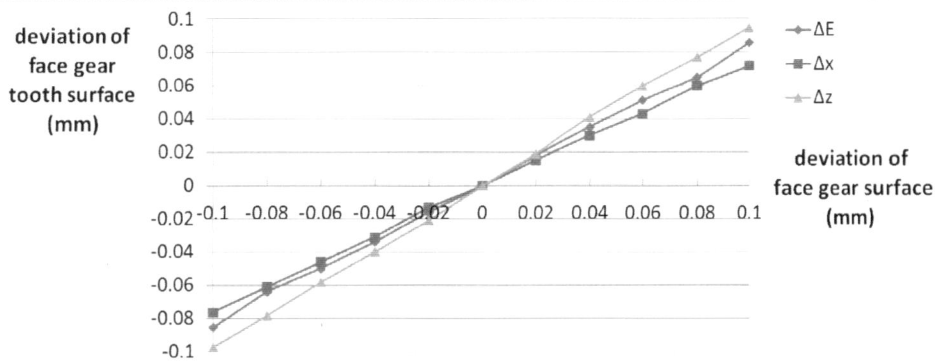

Figure 8. Relationship between deviation of face gear surface and deviation of grinding process

4. Conclusions

In this paper,relationship between deviation of grinding process and deviation of offset face gear surface have been studied.According to the generating theory of offset face gear, a grinding machine has been designed, and influences caused by deviation of offset distance ΔE and deviation of dressing process Δx,Δy and Δz have been analysed.Equations of offset face gear tooth surface containing deviaiton have been derived and figures of deviation cloud have been created with MATLAB,which provide a theoretical support for precise machining of offset face gear. The results show that Δz is the most influential deviation while Δy is the least one.

References

[1] Litvin F L, Fuentes A, Zanzi C, et al. Design, generation, and stress analysis of two versions of geometry of face-gear drives[J]. Mechanism & Machine Theory, 2002, 37(10):1179-1211.

[2] Litvin F L, Egelja A. Computerized design, generation and simulation of meshing of orthogonal offset face-gear drive with a spur involute pinion with localized bearing contact[J]. Mechanism & Machine Theory, 1998, 33(1-2):87-102.

[3] Handschuh R F, Lewicki D G, Bossler R B. Experimental testing of prototype face gears for

helicopter transmissions[J]. ARCHIVE: Proceedings of the Institution of Mechanical Engineers, Part G: Journal of Aerospace Engineering 1989-1996 (vols 203-210), 1994, 208(27):129-136.

[4] Tang J Y, Liu Y P. Loaded multi-tooth contact analysis and calculation for contact stress of face-gear drive with spur involute pinion[J]. Journal of Central South University, 2013, 20(2):354-362.

[5] Lewicki D G, Heath G F, Filler R R, et al. Face Gear Surface Durability Investigations[J]. Journal of the American Helicopter Society, 2008, 53(3):282.

[6] Hui G, Ning Z, Shengwen H. Tooth Deviation Analysis and Experimental Research of Face Gear Based on Disk Grinding Wheel[J]. Journal of Northwestern Polytechnical University, 2013.

[7] Peng X, Fang Z, Su J, et al. Theory analysis for application of disk tool of long radius in face gear machining[J]. Journal of Harbin Institute of Technology, 2013, 45(5):80-85.

[8] Wang Y, Lan Z, Hou L, ct al. A prccision generating grinding method for face gear using CBN wheel[J]. The International Journal of Advanced Manufacturing Technology, 2015, 79(9-12):1839-1848.

The Influence of Loading Force on Ultrasonic Drilling Garnet Ferrite

T H Pei[1,2], X F Zhang[2], Q Wang[1], T Y Zhang[1], Y Zhong[1] and Y Y Zhu[3]

[1] Aerospace Research Institute of Materials & Processing Technology. No. 1 South Dahongmen Road, Fengtai District, Beijing,100076, China
[2] Key Laboratory of Advanced Ceramics and Machining Technology, Ministry of Education, Tianjin University, Tianjin, 300072, China
[3] School of Mechanical Engineering, Beijin Institute of Technology, Beijing, 100081, China

xijiyu82@163.com

Abstract. In order to achieve high precision microholes of garnet ferrite, the rotary ultrasonic assisted machining and the tool load matching system were employed. An ultrasonic precise microholes technology experiment platform was built to machine garnet ferrite materials. A series of drilling experiments to machine garnet ferrite were presented in this paper, which were used to explore the influence of cutting force on tool wear. And the holes' exit aperture and wall surface roughness were applied to be evaluation parameters to explore the effect of loading force on machining quality. As a result, tool wear is aggravates as cutting force increases. And loading force shows different influence tendency on exit aperture size precision and wall surface roughness, the selection of proper loading force does good to control exit aperture size and hole wall surface roughness effictively.

1. Introduction

Garnet ferrite is a porous insulating composite composed of many grains and pores. It has excellent physical and chemical properties, such as high hardness, high electrical resistivity, high magnetic permeability, etc., which make garnet ferrite widely used in consumer equipment such as nuclear magnetic resonance, computer and other information equipment, and aerospace equipment such as aircraft and rockets. However, the unique structure makes it hard to process because of its high hard and brittle properties. Thus, How to machine it efficiently and accurately becomes an important research field[1].

In order to meet dimensional accuracy and assembly accuracy requirements, the ferrite product after fired needs to be machined. In view of the unique processing characteristics of ferrite materials, grinding is carried out generally. The surface topography of the parts obtained by grinding has t important impact on the performance of the parts[2]. Under the current technical conditions, the drilling experiment of microwave ferrite is usually processed by a tool for sintering or electroplating diamond abrasive grains. The processed microholes will have large inlet and outlet collapse, high hole wall roughness value, and make it hard to meet the expected requirements. To solve this problem, the cutting force should be controlled as low as possible during processing firstly. Secondly, the cutting force should be controlled as smooth as possible.

Ultrasonic-assisted machining is widely used in modern machining due to its low grinding force and good surface quality. However, traditional abrasive ultrasonic machining uses ultrasonic vibration of tools or workpieces to process hard and brittle materials through a suspension between the workpiece and cutting tool[3]. The material removal efficiency of this processing method is very low, at the same time, when the abrasive grains and the chip mixture return to the surface of the workpiece, the side walls of the processed hole wall will be worn again. This phenomenon is especially noticeable in microholes' processing[4]. Therefore, in order to solve the shortcomings of the traditional abrasive ultrasonic processing method, rotary ultrasonic assisted machining (RUAM) came into being. In RUAM, axial vibration together with tool rotation does the machining[5].Allowing for ferrite materials' high hardness, strong brittleness and too much stomatas which go against processing, it's necessary to apply RUAM into machining ferrite materials. Comparing with conventional drilling in previous studies, RUAM was used to drilling and polishing holes, low cutting force was provided and the machining accuracy was improved [6-7]. What's more, the smallest-diameter holes drilled by grinding can be obtained with Ultrasonic grinding[8].

RUAM provides less cutting force for precision hole making, but the cutting force during machining cannot be stabilized and is usually fluctuating. Therefore, in order to ensure the stable cutting force and improve the machining quality during the machining process, the load matching system is applied to the machining process. J F Zhang and others of Chongqing University designed a new type of ultrasonic pressurizing device during the study of ultrasonic machining principle and process parameter optimization of brittle materials. In the ultrasonic machining process, the lifting movement of the pontoon can be controlled by the liquid level which means relatively stable load[8]. G Ya, et al. [9-10] used a mechanical device such as a spring or a weight to establish a tool load matching system. Based on the theory of hydraulic loading, P. Legge established a load matching system with an ultrasonic resonance frequency of 20 kHz, drilling glass and crystalline alumina, respectively, with dimensional accuracy of 0.0254 mm and 0.0127 mm[11].

Ultrasonic machining technology and tool load matching system are widely used in various machining fields, and load force also has an important impact on machining quality. However, the effect of load force on tool wear and machining accuracy has rarely been reported in papers. Therefore, a drilling experiment research was carried out, garnet ferrite was used as the workpiece, and air-floating guide rail and cylinders were used to achieve precise load force to explore the influence of load force on tool wear. In the process of analysis, the outlet diameter and surface roughness of the hole were used as indicators to evaluate the parameters to study the effect of different load forces on machining accuracy.

2. Experimental conditions and schemes

2.1 Material and cutting tools

The experimental specimens are composed of fully sintered garnet ferrite (molecular formula $R_3Fe_5O_{12}$). The cutting tools are electroplated diamond tools. The related parameters of the experimental materials and cutting tools are shown in Table 1.

Table 1 Relevant parameters of cutting tools and experimental materials

Tool substrate	Material	Stainless steel
	Diameter	0.60 mm
Tool abrasive	Material	Diamond
	Granularity	200#
workpiece	Material	Garnet ferrite
	Dimension	30×25×4 mm^3
	Density	3.69×103 kg/m^3

2.2 Experimental and measuring devices

The ultrasonic vibration system consists of an ultrasonic generator, a power amplifier and an Ultrasonic Processing Head. As shown in Figure 1, the ultrasonic processing head can be mounted on the spindle of the machine to clamp tools of different sizes, and the ultrasonic electrical signal can be transformed to be ultrasonic mechanical vibration of the cutting tool.

Figure 1. Ultrasonic Processing Head

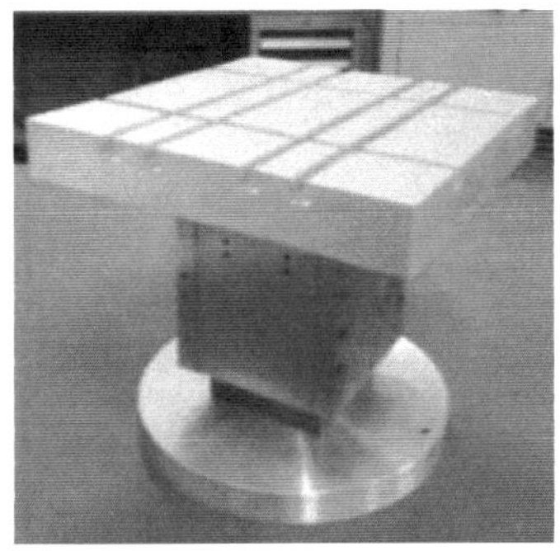

Figure 2. Pneumatic Flotation Workbench

The load matching system is composed of a software system, an air supply system and a Pneumatic Flotation Workbench(PFW). The software system is able to control and monitor all of the whole system. The air supply system is composed of lubricating gas path and supporting gas path. The air floating table is the execution element of the load matching system, as shown in figure 1-2, which can maintain a constant the loading force. The movement principle of air floating table can be explained by the resultant force of air floating table, as shown in Formula 1-1.

$$F_R = P \cdot S - G - F \quad (1)$$

F_R represents the resultant force on the moving guide rail of PFW. P is the stable air pressure at the top of the cylinder. S is the effective area of the cylinder. G means the whole weight of the moving guide rail, fixture and workpiece of PFW. Finally, F represents the carrying capacity between the cutter and the workpiece.

The Ultrasonic Processing Head and load matching system invented by authors can be directly connected to the ordinary machine spindle. In this paper, XH716D HanChuan CNC machine was selected as the test machine. The ultrasonic processing head was connected to machine spindle. PFW was directly installed on the working platform of CNC machine.

After machining microholes, The WE6800-3 high precision microscopy was used to observe the diameter. The nanovea ST400 profile meter was employed to test the hole wall line roughness. In order to collect reliable data, each test was repeated three times and the average value was obtained by results.

2.3. Experimental scheme

2.3.1 The impact of loading force on tool wear. In this experiment, microholes was drilled by ultrasonic processing head and load matching system. The hole diameter is Φ0.65mm,each cutting tool drilled 4 holes, The loading force F was set as 5N, 10N, 16N, 29N and 35N, and the other processing parameters were feed speed Vw=1.0mm/min, tool linear speed Vs=0.191m/s, and ultrasonic power P=40% (ultrasonic maximum power 25W).

2.3.2 The impact of loading force on processing quality. Based on the previous technical accumulation in the laboratory, taking the interaction of different machining parameters into account, three groups of optimized machining parameters were selected for carrying out optimization experiment. Feed speed Vw, cutter linear speed Vs, and ultrasonic power P are shown in **Table. 2**. To drill different microhole separately under the above three groups of processing parameters, the loading force were set to 5N, 10N, 16N, 29 N, 35N. The theoretical diameter is Φ 0.65 mm. The diameter and surface roughness data were used to be the evaluation index of processing parameters, which had been recorded.

Table 2 Processing parameters

Group	Feed speed *Vw* (MM/min)	Tool linear speed *Vs* (M/S)	Ultrasonic power *P* (%)
1	1.0	0.191	40
2	1.5	0.157	40
3	1.5	0.225	30

3. Experimental results and analysis

3.1 The impact of loading force on tool wear

Under different load conditions, the tool wear state is as shown in **Figure 3a.** to **Figure 3f.**:

a) Tool wear state before machining

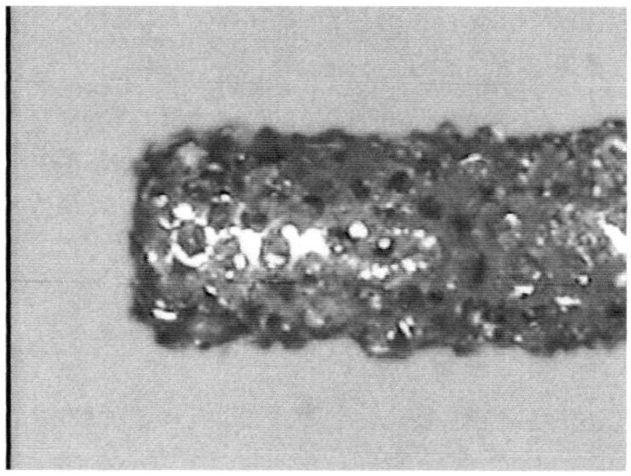

b) Tool wear state under 5N

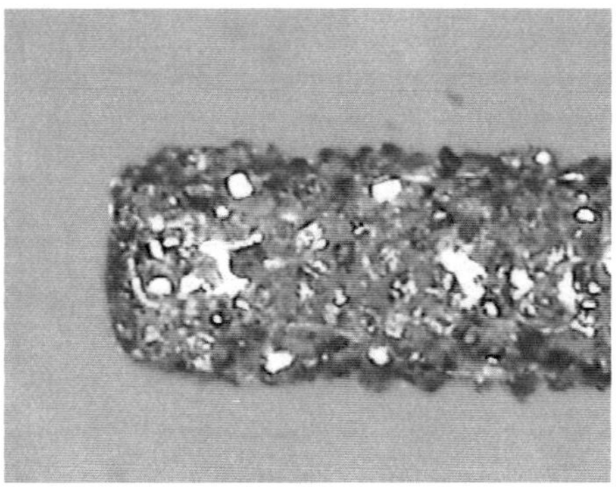

c) Tool wear state under 10N

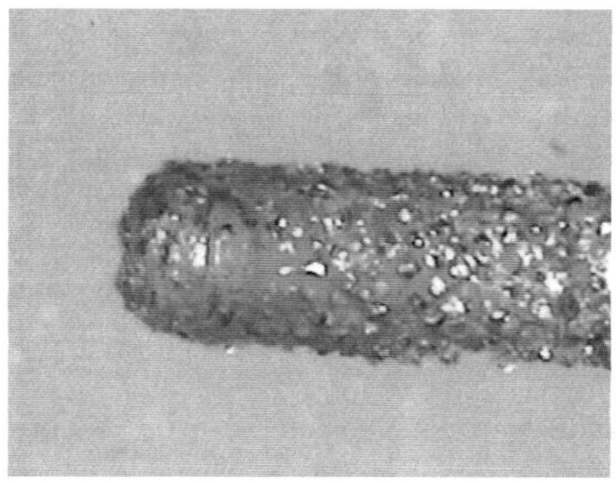

d) Tool wear state under 16N

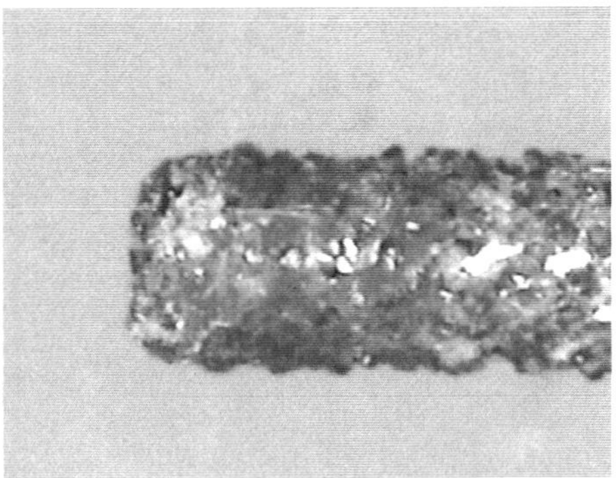

e) Tool wear state under 29N

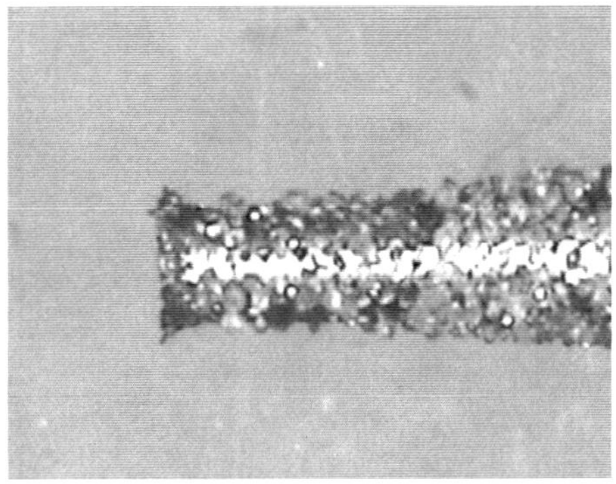

f) Tool wear state under 35N

Figure 3. The Tool wear state

Focus on **Figure 3a - f,** it can be found as followed. There was a small amount of diamond grits wear when the loading force set to 5N and 10N and a large number of grits dropped down when loading force was 16 N. Binder began to fall off when loading force was 29N, and the base of cutting tool had been involved in processing. When the loading force was set to 35N, cutting tools began to fracture at the variable cross-section (effective length). Consequently, when other parameters didn't changed, with the increase of the load, tool wear became serious. This is because with the increase of the loading force, the tool grinding effect was enhanced, the impact force was increased, at the same time, the size of the chip became larger, material removal rate increased. All of this means, chip removal became more difficult, and the gap between the tool and the workpiece got smaller, so the tool wear became faster.

It was also found that the most worn area was the tool neck. Combined with the mechanism of drilling ultra-precision hole, considering the sufficient supply of cooling fluid for this experiment, the effect of cooling fluid would not be taken into account here. The analysis showed that in the machining process, due to the improvement of the loading force, the cutting efficiency was significantly improved, and the tool abrasive passivation was intensified. At this time, the chip was hard to be removed, and would accumulate in the back of the tool head. With rapid rotation of the cutting tool, there was a serious erosion caused by the abrasive particle and matrix of tools. All of this caused the neck destroyed more and more seriously with the improvement of loading force.

3.2 The impact of loading force on exit aperture diameter

In the process of machining garnet ferrite, the amplitude range of the cutting force is 4.5N-18N. Thus, the loading force is set to 5N, 10N, 16N and 29N. The experimental results of diameter are recorded in **Table 3.**:

Table 3. Diameter results

Unit: mm

Group	Loading force F			
	5N	10N	16N	29N
1	0.667	0.724	0.798	0.682
2	0.664	0.757	0.734	0.698
3	0.668	0.753	0.726	0.694

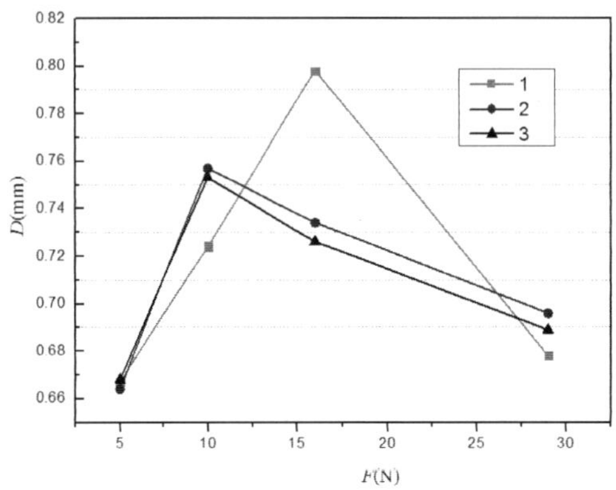

Figure 4. The impact of loading force on exit aperture diameter

In **Figure 4.** with the gradual increase of the loading force, the variation trend of pore size under three experimental parameters increased firstly and decreased in the following time. It could be anticipated that the action of drilling microholes combined the action of the cutter and the air floating worktable. With the increase of the loading force, the influence of cutting on the workpiece was strengthened. However, when the load exceeds the critical point, either way, when the loading force exceeded the maximum cutting force in the traditional drilling, the tool wear becomes quite acute, and the tool diameter would decreased significantly (as shown in **Figure 3b-3f.**). Therefore, although the damage to the surface of the microhole would trend to get serious and the chip may even increase when the tool is worn, the outlet diameter of the microhole will obviously decrease.

In addition, the diameter of the microhole fluctuated between 0.66 and 0.80mm, which showed that when the constant loading force was set within the change range of conventional machining cutting force, the diameter of the microhole would change in different degrees. When the loading forcce exceeded the range of the conventional machining force, the outlet diameter of microhole decreased, but the cost was the sharp wear of the tool. Therefore, for garnet ferrite materials, the loading force should be controlled at about 5N, when a good inlet diameter is obtained.

3.3 The impact of loading force on wall surface roughness

Table 4. Surface roughness results

Unit: μm

Group	Loading force F			
	5N	10N	16N	29N
1	1.832	2.206	1.867	2.105
2	1.647	1.825	2.769	3.552
3	2.061	2.113	1.839	2.233

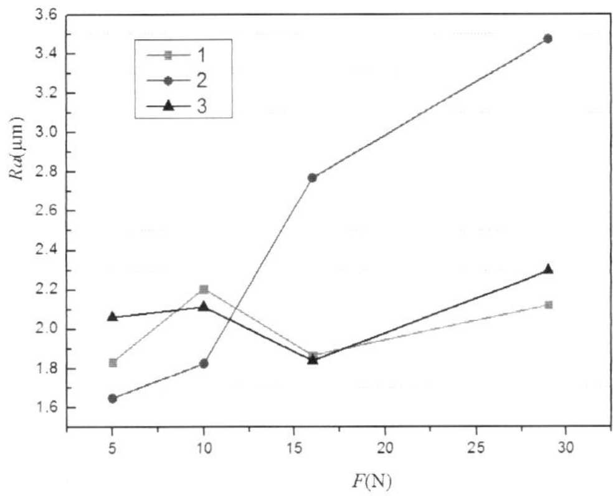

Figure 5. The impact of loading force on surface roughness

As shown in **Table 4.**,under the loading force of 5N, 10N, 16N and 29N, the average roughness of the wall surface is 1.847 m, 2.048 m, 2.158 m and 2.630 m. Compared with **Figure 3b-3f**, it was shown that the surface roughness of the hole wall increased with the increase of the loading force. This is because the impact effect of ultrasonic tools was enhanced along with the increase of the loading force, caused in a large amount of debris. At the same time, the impact effect led to the fracture of the bond of the grinding wheel, and the grinding wheel's abrasive particles would collapse.

According to **Figure 5**, the surface roughness value increased gradually due to the effect of larger debris and cutter grinding grains on the ploughing and sliding friction of the inner wall of microhole. In addition, when the loading force is 16N and below, the fluctuation of PFW promoted the removal of chips to a certain extent. When the loading force was 29N, the cutter and the workpiece contact for a long time relatively, the PFW would not float up and down, the chip was hard to be removed away. when the loading force was 29N, the roughness was the worst, and the change range was largest. Therefore, the smaller of the loading force is, the better wall surface quality will obtain. In order to obtain smaller pore wall surface roughness, the loading force should be controlled at about 5N.

4. Conclusion

This paper carried out different processing experiments on garnet ferrite, and analyzed and explored the influence of load on tool wear, microhole exit diameter and micro-hole wall surface roughness in detail. The conclusions are as follows:

(1) The loading match system can be used to output different loading force to process the workpiece. During drilling, with the increase of tool loading force, tool wear would become more and more serious, and the tool neck was the most serious area of tool wear. In order to improve the tool life and machining precision, the structure and material of the tool should be designed reasonably in actual machining.

(2) The diameter of the microhole increases firstly and decreases after that. When the loading force is controlled within the floating range of the cutting force in traditional machining, the dimensional accuracy of the aperture is poor. In order to obtain the best aperture, the actual machining should be set to 5N.

(3) The surface roughness of microhole wall increases with the increase of the loading force. And when the loading force exceeds the maximum cutting force in traditional machining, the wall surface roughness will be seriously increased. In order to obtain smaller surface roughness of the hole wall, the loading force should be strictly controlled at 5N.

Acknowledgments

The authors wish to thank the CALT Process Improvement & Research Projects in 2019 (Grant No. GYGJ19-703-18) and the National Natural Science foundation of China Project(Grant No. 51205285).

References

[1] Z M Wang 2013 *Ferrite Manufacturing Process Technology*(Chongqing: Chongqing University Press) p 1-9

[2] S Y Yu and B Lin 2008 *The machining technology and application of Engineering ceramic* (Beijing: China Machine Press) p 180-193, 207-213

[3] B X Jia, W F Bian and W. S. Zhao 2007 The key technology of Ultrasonic machining micro-holes *J. Mech. Engi.* 43(11), p 212-216

[4] JunWei L, Dae Kyun Baek, Tae Jo Ko 2014 Chipping minimization in drilling ceramic materials with rotary ultrasonic machining *Int J. Adv Manu. Tech.*, 72, p1527-1535

[5] Takuya Asami, Hikaru Miura 2012 Characteristics of Hole Machining of Brittle Material by Ultrasonic Complex Vibration *Jap. J. App. Phys.*, 51, p07GE07.

[6] J Pujana, A Rivero, A Celaya, L N Lo´pez de Lacalle 2009 Analysis of ultrasonic-assisted drilling of Ti6Al4V *Int. J. Mach. Tools & Manu*, 49, p500-508.

[7] Kai Egashira∗, et al. 2014 Drilling of microholes down to 10m in diameter using ultrasonic grinding *Prec Eng*, 38, p605-610.

[8] J F Zhang, L Zeng 1993 Study on ultrasonic machining of non-metallic hard and brittle materials. *New Tech. & New Proc,* 06, p 12-13

[9] G Zha, H W Qin, Y W Xu, et.al 2000 The experimental research of rotary ultrasonic machining *Avia. Manu. Tech.* ,06, p56-59

[10] X L WANG, J H Zhang, Q H Zhang, et.al 2009 Study on surface temperature field of Ultrasonic assisted End-Grinding *J. Wuhan Univ. Tech.* 03, p 541-544

[11] P. Legge 1966 Machining without abrasive slurry *Ultrasonic*, 4(7), p 157-162

IOP Conf. Series: Materials Science and Engineering **647** (2019) 012017 doi:10.1088/1757-899X/647/1/012017

Performance Analysis of by-pass Excitation Cable Force Sensor

Zhou Yishu[1], Zhang Jinning[1,*], Liu Chang[1] and Cao Jinjin[1]

[1]Research Institute of Highway Ministry of Transport No. 8 Xitucheng Road, Haidian District, Beijing

[*]Corresponding Author E-mail: jinning.zhang@foxmail.com

Abstract. The sleeve cable force sensor based on magnetoelastic effect has developed rapidly in recent years, and has been widely used in the field of bridge cable force measurement in China. Existing sleeve cable force sensor needs to be installed at the beginning of construction, so it is difficult to reinstall them later, and its long-term performance is also difficult to evaluate and measure online. To solve this problem, the tension measurement of the designed by-pass excitation cable force sensor is carried out, and the influence of signal line length and ambient temperature on the measurement results is analyzed. The results show that the designed by-pass excitation cable force sensor has high linearity, which is expected to solve the problem of the existing sleeve cable force sensor.
Keywords: cable force measurement, bypass excitation, magnetoelastic effect, cable force sensor

1. Introduction

With the continuous progress of bridge technology in China, cable technology is increasingly widely used in long-span bridges. Cable is the most important part of the whole structure as load-bearing components of various large-scale buildings and structures, and its stress condition plays an extremely important role in the overall structural safety [1, 2]. Because of long-term alternating stress, corrosion and wind-induced vibration environment, it is easy to cause local fatigue and damage of cables, which not only shortens the service life of cables, but also directly affects the internal force distribution and structural linetype, endangering the safety of the whole structure. Real-time measurement of cable force is the main means to judge whether the bridge structure is safe or not. At present, many measurement methods have been developed, such as vibrating string type, optical fiber strain type, pressure gauge type, magnetic flux type, etc. [3-7]. Among them, the cable force sensor based on magnetic flux method is widely used in practical engineering because of its excellent sensing performance and long-term stability [8-10].

The existing magnetic flux cable force sensors are designed with sleeve type, which has simple structure, uniform magnetic field, good linearity and accurate measurement results. However, due to the sleeve design, it is necessary to be installed at the initial stage of bridge construction. Once completed, it is difficult to carry out subsequent disassembly and assembly. After long-term operation, the measurement accuracy of the sensor cannot be determined, and it is difficult to disassemble and calibrate. Therefore, a kind of high-precision cable force sensor is urgently needed, which is easy to disassemble and assemble later. This kind of sensor can be used for completed bridges and to calibrate existing cable force sensors in situ. In this paper, a kind of by-pass excitation cable force sensor which is convenient for later disassembly and assembly is introduced. The tension calibration and measurement experiments are carried out, and some factors affecting the measurement accuracy are analyzed experimentally.

2. By-pass Excitation Cable Force Sensor

In this paper, experimental verification and performance analysis of the designed by-pass excitation cable force sensor are carried out. The by-pass excitation cable force sensor is mainly composed of bypass core, cover plate, excitation coil and measuring coil, and its structure design is shown in Figure 1. The bypass core and cover plate are processed with pure electrical iron to ensure good magnetic conductivity, and the system magnetic circuit is composed of the tested cable, the bypass core and cover plate. When the sensor is connected, it is necessary to ensure that the magnetic field generated by the four excitation coils is in the same direction when passing through the cable. The sensor is arranged on both sides of the cable in a symmetrical structure. Figure 2 shows the installation effect of the sensor.

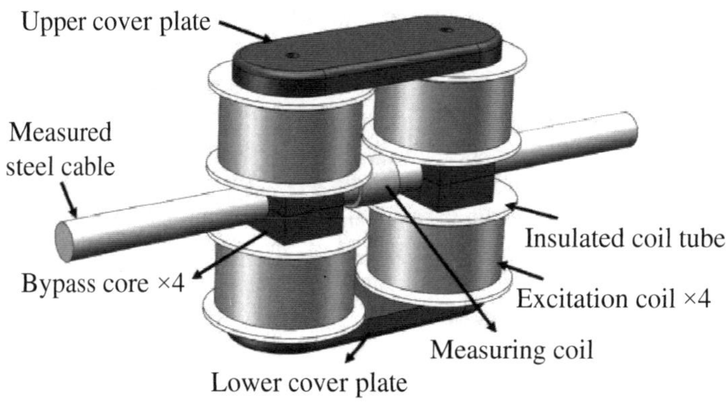

Figure 1. Structural design of bypass excitation sensor

Figure 2. Installation effect of bypass excitation sensor

3. Sensor performance test

In order to test the performance of the designed sensor, it needs to be connected to the main machine of magnetic flux cable force measurement, namely the magnetoelastic instrument (OVM PowerStress EM Sensor Measurement Unit). In the experiment, the tension of the cable is exerted by a tension machine, and then measured by a magnetoelastic instrument. The experimental system is shown in Figure 3.

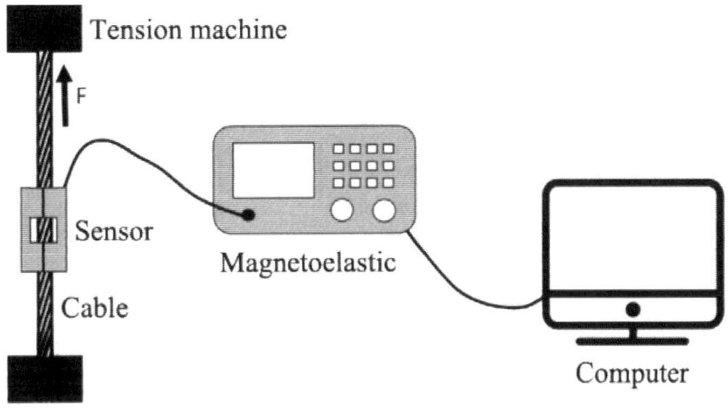

Figure 3. Cable force measurement experimental system

3.1 Tension measurement experiment

In the experiment, the induced integral voltage was measured five times under each tension, and the average value was taken. The measured data and calculated average value are shown in Table 1. The average value was linearly fitted and the results are shown in Figure 4. It can be seen from the chart that the fitting linearity of the tension and the induced integral voltage is 0.99854, which is close to the level of the existing sleeve magnetic flux sensor. Since the parameters of sleeve cable force sensor are used in the experiment, these parameters have not been optimized for the by-pass cable force sensor. After a large number of experiments and parameters optimization, the measurement sensitivity and accuracy of the by-pass cable force sensor may be further improved.

Table 1. Measurement data of bypass excitation sensor

Tension (KN)	Measured integral voltage (V)					Average integral voltage (V)
	First	Second	Third	Fourth	Fifth	
4	66.1	66.1	66	66.1	66.2	66.1
32	70.6	70.6	70.5	70.5	70.6	70.56
61	75.6	75.5	75.2	75.4	75.4	75.42
90.5	80.1	79.8	80.1	79.9	80	79.98
121.1	85	84.8	84.3	84.5	84.5	84.62
150.3	88.6	88.3	88.9	88.4	88.5	88.54
179.8	93	93	93.1	92.5	92.4	92.8

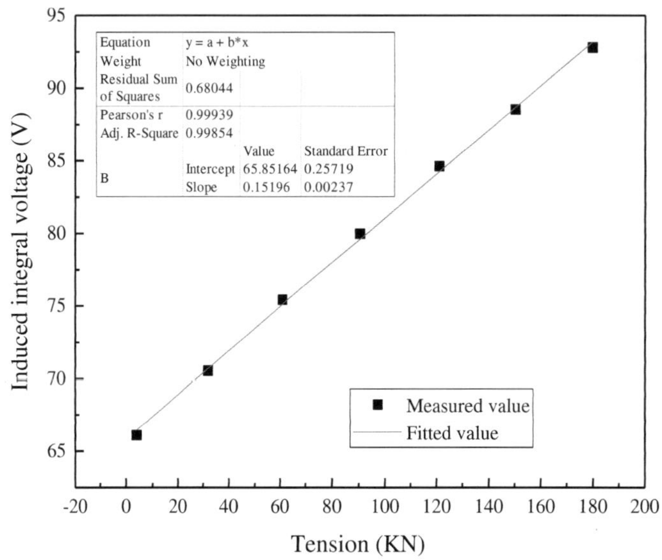

Figure 4. Measurement fitting curve of bypass excitation sensor

3.2 Influence analysis of signal line length

In field measurement, multiple sensors often share a set of measuring instruments. There will be a long signal line between the measuring instrument and the sensor, which will adversely affect the signal transmission and the measurement results. In order to understand the influence of the length of signal line on the experimental results, the tension in signal lines of different lengths was measured at the standard tension of 120 KN. The measurement data and error analysis are shown in Table 2.

Table 2. Measurement results of tension in signal lines of different lengths

Length of signal line (m)	Standard tension (KN)	Measured average tension (KN)	Difference value	Same level error	Comprehensive error
52	120.81	116.67	-4.14	-3.43%	-2.30%
105	120.67	113.50	-7.17	-5.94%	-3.98%
157	120.54	115.07	-5.47	-4.54%	-3.04%
210	124.30	115.40	-8.90	-7.16%	-4.94%

It can be seen from the experimental results that the length of the signal line does affect the measurement results. Therefore, different line lengths should be calibrated or compensated in practical application.

3.3 Influence analysis of temperature

The permeability of the material is a physical quantity that varies with temperature, and the temperature change will have a significant impact on the measurement results. In actual product development, sensor calibration experiments at different temperatures are needed. In the actual measurement, it can be used as a temperature compensation database to correct the cable force data at the actual temperature.

The bypass excitation sensor designed in this paper also needs large-scale temperature calibration experiments for product development. This paper mainly introduces the principle experiments and results of design performance, so only the permeability analysis at different temperatures is carried out.

In the experiment, the cable and the by-pass cable force sensor installed on it were placed in the oven. The oven temperature can be read directly. At the same time, a temperature sensor was placed on the cable force sensor to measure the temperature of the cable force sensor accessories.

Table 3. Measurement data of bypass excitation sensor at different temperatures

Oven display temperature (°C)	Magnetoelastic temperature (°C)	Measured integral voltage (V)			Average integral voltage (V)	Permeability
		First	Second	Third		
12	12	89.40	89.50	89.50	89.47	1.437920
25	23	88.30	88.30	88.30	88.30	1.399981
31	30	87.40	87.40	87.40	87.40	1.370714
42	42	85.80	85.80	85.70	85.77	1.317600
50	50	84.10	84.20	84.20	84.17	1.265570

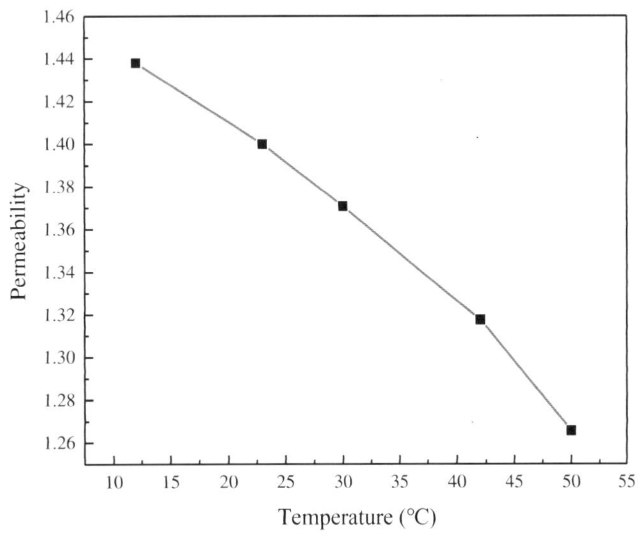

Figure 5. The relationship between permeability and temperature

Table 3 shows the experimental results of the bypass cable force sensor at different temperatures. Figure 5 shows the relationship between permeability and temperature. It can be seen from the above measurement results that the permeability decreases with the increase of temperature, which is in line with the theoretical expectation.

4. Conclusions

In this paper, a kind of by-pass excitation cable force sensor is introduced, and the measurement performance of the sensor is tested. The results show that the by-pass excitation cable force sensor has high linearity, which ensures high measurement accuracy, the ambient temperature and the length of signal line have an impact on the measurement results, which must be considered in practical applications. The by-pass excitation cable force sensor is convenient to install in the measurement site, which can solve the problem that sleeve cable force sensor cannot be installed subsequently, and it is expected to perform effective on-site calibration of the existing sleeve cable force sensor.

References

[1] Wang M L, Wang G. Measurement of stresses of large steel cables [M]. 2003.
[2] Wang G. The Development of a Cable Force Detecting System based on the Magneto Elastic Effect [D]. Xi'an: Chang'an University. 2017.
[3] Park S H, Kim J W, Nam M J, et al. Magnetic Flux Leakage Sensing-Based Steel Cable NDE Technique Incorporated on a Cable Climbing Robot for Bridge Structures [J]. Advances in Science & Technology, 2013, 83:217-222.
[4] Zhou X F, Xu J F, Liang L. Design and Analysis of Fiber Bragg Grating Cable Tension Sensor [J]. Chinese Journal of Sensors and Actuators, 2007, 20(5): 1007-1011.
[5] Jiang X, Jing Y Z, Zhang H C. Vibrating Wire Sensor based on Sweeping Frequency to Excite Single Coil [J]. Transducer and Microsystem Technologies, 2001, 20(5): 22-4.
[6] Park S, Kim J W, Moon D J. NNon-contact Main Cable NDE Technique for Suspension Bridge using Magnetic Flux-based BH Loop Measurements[C]. International Society for Optics and Photonics, 2015, 9437: 94370M.
[7] Wu K X, Liu K M, Yang J X. Measuring System of Cable Tension Based on Frequency Method [J]. China Journal of Highway and Transport, 2006, 19(2): 62-6.
[8] Cappello C, Zonta D, Laasri H A, et al. Calibration of Elasto-Magnetic Sensors on In-Service Cable-Stayed Bridges for Stress Monitoring [J]. Sensors, 2018, 18(2): 466.
[9] Deng N C, Long Y, Sun L M, et al. Magnetic Flux Sensor and its Application in Bridge Engineering [J]. Prestress Technology, 2008, 02: 17-20.
[10] Lenz J, Edelstein A S. Magnetic Sensors and Their Applications [J]. IEEE Sensors journal, 2006, 6(3): 631-649.

AUTHOR INDEX

Institute of Physics
Dirac House, Temple Back
Bristol BS1 6BE UK

ISSN: 1757-8981
ISBN 978-1-7138-0908-1

9 781713 809081